THE CHILD'S CONSTRUCTION
OF QUANTITIES

BOOKS BY JEAN PIAGET

The Child's Conception of Movement and Speed
The Child's Conception of Number
The Child's Conception of Physical Causality
The Child's Conception of Time
The Child's Conception of the World
The Child's Construction of Reality
Insights and Illusions of Philosophy
Judgment and Reasoning in the Child
The Language and Thought of the Child
The Mechanisms of Perception
The Moral Judgment of the Child
The Origin of Intelligence in the Child
Play, Dreams and Imitation in Childhood
The Principles of Genetic Epistemology
The Psychology of Intelligence
Structuralism

With Bärbel Inhelder

The Child's Conception of Space
The Early Growth of Logic in the Child: Classification and Seriation
The Growth of Logical Thinking: From Childhood to Adolescence
Mental Imagery in the Child
The Psychology of the Child
Memory and Intelligence

With Bärbel Inhelder and Alina Szeminska

The Child's Conception of Geometry

With Paul Fraisse (eds)

EXPERIMENTAL PSYCHOLOGY: ITS SCOPE AND METHOD
History and Method
Intelligence
Learning and Memory
Motivation, Emotion and Personality

The Child's Construction of Quantities

Conservation and Atomism

JEAN PIAGET

and

BÄRBEL INHELDER

Translated from the French by
ARNOLD J. POMERANS

LONDON
ROUTLEDGE & KEGAN PAUL

First published in 1974
by Routledge & Kegan Paul Ltd
Broadway House, 68–74 Carter Lane,
London EC4V 5EL
Set in Monophoto at Thomson Press (India) Limited
and printed in Great Britain by Lowe & Brydone
(Printers) Ltd., Thetford, Norfolk

ISBN 0 7100 7887 0

Contents

Preface   vii

PART I
CONSERVATION

1 The Conservation of Matter and Deformations of a Ball of Modelling Clay   3
2 The Conservation of Weight upon Deformations of a Clay Ball   22
3 The Conservation of Volume at Equal Concentrations of Matter   47

PART II
FROM CONSERVATION TO ATOMISM

4 The Destruction of Matter and the Dissolution of Sugar   67
5 The Conservation of the Sugar and the Beginning of Atomism   81
6 The Conservation of the Weight and Volume of the Dissolved Sugar and the Completion of Atomism   98

PART III
COMPRESSION, DECOMPRESSION AND DENSITY

7 The Expansion of a Maize Seed and of a Column of Mercury   119
8 Differences in Density   137
9 Special Problems Posed by the Relationship between Weight and Quantity of Matter   154

PART IV
FORMAL COMPOSITIONS

10 The Composition of Asymmetrical Relations and Differences in Weight   183
11 Simple and Additive Compositions of Equivalent Weights   203

12 Simple and Additive Compositions of Equivalent Volumes
and the Discovery of the Displacement Law   232

Conclusion   269

Index   281

Preface

In an earlier work[1] we have tried to show how children construct numerical concepts by fitting the logical multiplication of classes and relations into one operational whole. As a result, they are able to extend logical or intensive quantities (characterized by a simple relationship between the parts and the whole) into a numerical or metric quantification based on the construction and iteration of units.

Now, there is one quantity that is even more general than number or measurement, and includes them as special cases, namely the extensive quantity characterized by the comparison of the parts without the specification of a unit.

It is the construction of this type of continuous quantity which we shall be examining in the present work. To that end we have thought it best to confine ourselves to problems of physical quantification, so much so that the title of this book should perhaps have been the Child's Construction of Material Quantities. In it, we shall not be dealing with such fundamental quantities as time, speed, and space, to which we hope to return in subsequent studies, but shall merely try to discover how the child, once he has grasped the elementary notions of logical and numerical quantity, manages to generalize them and to apply them to the main material data he encounters: quantity of matter, weight and physical volume.

Though this problem may appear limited in scope, in fact, it raises the very broad question of the quantification of qualities in general and, consequently, that of the intellectual organization of the external world.

To begin with, the quantification of qualities is something quite other than the construction of number. Numerical concepts appear in connection with discrete objects, which are combined into classes according to their equivalences, seriated according to their differences, or classified and seriated by the fusion of these two processes, which is tantamount to arranging them in similar units and hence to counting them off. A quality, by contrast, is a continuum that does not appear in the form of objects but as an irreducible characteristic of things, apprehended thanks to particular actions

[1] J. Piaget and A. Szeminska, *The Child's Conception of Number*, Routledge & Kegan Paul, 1952.

on the part of the subject. Substantiality is therefore the central characteristic of what can be lifted up, and voluminosity of what can be enclosed or contained. Our first problem is to determine how the child constructs extensive or even measurable quantities from these, originally phenomenalistic and egocentric, qualities.

Now, in trying to solve that problem, we were immediately faced with another, namely the conservation of matter, weight and volume (at equal concentrations), because it is quite clear that conservation is both a condition and also a result of quantification. That being the case, we want to know how the child arrives at the principles of conservation. Is he guided by experience, or by mental constructions of his own? Moreover, we are also forced to examine another question, closely bound up with the last, namely that of atomism as the instrument of conservation and quantification. And atomism, no less than the discovery of quantitative invariants, calls for a discussion of the role of experimental data and of deduction in the organization of the universe of physical quantities.

Studying the quantification of qualities through the construction of the physical principles of conservation and through the child's gradual and spontaneous elaboration of atomism raises the wider problem of the relations between the mind and objects, or rather of the interactions between mental activity and experience.

If we quantify a quality we have to measure it sooner or later. Now, the only way of measuring a substance is through its weight, volume and density (and subsequently through its mass, etc.). And when we measure a weight we quickly learn to distrust our muscular impressions and come to rely instead on the balance; similarly, when we measure a volume we sooner or later cease to rely on visual impressions and on touch, and learn to apply such indirect methods as water displacement. Now, as soon as we use measuring instruments or measurements based on physical relations we begin to frame laws: we cease to rely on reason and deduction alone, and begin to organize experimental induction. That is why we shall devote the last chapter of this book to the child's gradual elaboration of the law on which he bases his quantifications.

The study we are about to present, far from being self-contained, opens up several new perspectives. We shall undoubtedly have to devote a special volume to the child's inductive discovery of experimental laws in general, not merely in the laboratory but in his daily contacts with the external world. We can only hope that circumstances will allow us to bring this new enterprise to a successful conclusion.

JEAN PIAGET
BÄRBEL INHELDER

December 1941

PART I
Conservation

CHAPTER ONE

The Conservation of Matter and Deformations of a Ball of Modelling Clay

Even when he has mastered what is unquestionably the first of the principles of conservation, namely the invariance of the shape and dimensions of solid objects, the child still has to resolve a number of other problems posed by the conservation of matter. Thus if the object of perception recedes or approaches, its invariance must be deduced from the co-ordination of successive perceptions; but if it has been subjected to real transformations (e.g. division or re-arrangement of the parts) under the child's eyes, he must decide whether these changes affect the total volume of the object, its weight or quantity of matter, or merely affect its geometrical aspects (shape and dimensions).

These problems are, of course, much more difficult to solve than that of the conservation of the object as such. Moreover, while the invariance of solid objects is acquired during the sensori-motor stage (beginning at the end of the first year of life), the conservation of matter, weight and volume is not constructed until a later phase of development (generally between the ages of seven and twelve years), and this because it calls for the disassociation of the different quantifiable aspects of matter (weight, volume, etc.) and also for the quantification of these qualities. No wonder then that, between the conservation of the object and that of the quantifiable elements of matter, there should appear a host of other constructions which must be completed before the quantification of physical quantities can be tackled. They include particularly the elementary logical and arithmetical notions which we have discussed elsewhere,[1] and this because the types of conservation we are about to examine are direct extensions of these notions. Moreover, the operations involved are the same, so that the conservation of the quantity of matter— which we shall simply call the 'conservation of substance or of matter' and which is the starting point of the quantifica-

[1] J. Piaget and A. Szeminska, *The Child's Conception of Number*, Routledge & Kegan Paul, 1952.

3

tion of such physical qualities as weight, volume, etc. – can also be considered the end points of the elementary mathematization thanks to which numbers are generated. We shall therefore resume the study of quantification where we left it when dealing with arithmetic, though here we shall be dwelling on the intellectual conquest of physical reality and no longer on quantifying operations as such.

§1. *Method and general results*

The method we shall be using in the first part (Chapters 1–3) is extremely simple. The child is handed a ball of modelling clay, together with a lump of the same material, from which he is asked to make another ball 'as big and as heavy as the first'. Once he is satisfied that the two balls are identical, the demonstrator changes the shape of one of them by drawing it out into a coil (a roll or sausage shape), by flattening it into a disc, or by cutting it up into pieces. He asks the child if the two objects still have the same weight, quantity of matter, volume, etc. The child is expected to justify all his answers, so that it is possible to determine not only whether or not he accepts the idea of conservation but also how he substantiates and elaborates it.

This method has shown us that the elaboration of conservation passes through three stages whose chronological appearance suggests that they must represent three distinct stages of intellectual development. Thus, at a point, the child becomes convinced that a change in the shape of the ball produces no changes in its weight, when previously he was certain that the weight varies with each deformation. This invariance is grasped at about the age of ten or eleven years on average, and is not the first to appear. As Kant has put it, the assertion that all matter has weight is a synthetic judgment, because the concept of weight has no analytic links with that of matter itself. If even the physicist has some difficulty in understanding this distinction, the young child cannot possibly grasp it, which explains why seven- to eight-year-olds, whilst continuing to question the conservation of the weight of the deformed balls, nevertheless maintain that their quantity of matter remains constant. The child expresses the situation by saying that it is 'the same clay as before' but no longer 'the same weight'. In other words, he quantifies a kind of undifferentiated quality which one might call the substance, before he can quantify those special qualities of weight or of volume which constitute its attributes. Or, to put it differently, he arrives at the idea of a global quantity before he can construct such differentiated quantities as weights or volumes. Nor does he equate quantity of matter with volume; only at about

4

the age of eleven or twelve, i.e. after the discovery of the conservation of weight, does he begin to realize that a clay ball immersed in a glass of water will displace its own volume of water, no matter how its shape is altered.

The clay-ball experiment, as we have said, reveals the successive nature of the elaboration of the principles of the conservation of matter, weight and volume. In this development, we can distinguish the four major stages we shall be analysing throughout this book. During the first of these (up to about the age of seven or eight on average) the child does not grasp the conservation of substance, or of weight, or of volume. During the second (from the age of about eight to ten years) he grasps the conservation of substance but not those of weight and volume. During the third (from the age of ten to between eleven and twelve) he grasps the conservation of matter and weight but not yet that of volume. Finally, during the fourth stage (starting from eleven or twelve years) he grasps all three types of conservation, and tends to reduce the idea of substance to those of weight and volume. Moreover, Stages II–IV can each be divided into two sub-stages, the first involving a number of intermediate reactions, and the second more typical of the actual stage. Thus, in the present chapter, we shall see that it is possible to distinguish between Sub-stage IIA, in which the conservation of substance is accepted in certain cases and not in others; and a Sub-stage IIB, when it is accepted under all circumstances. Moreover, it is worth stressing that the three stages we have been able to distinguish in respect of the conservation of matter (Stage I, Sub-stage IIA and Sub-stage IIB) correspond to the three stages we have distinguished in *The Child's Conception of Number*.[1]

§2. *The first stage: absence of conservation*

The first stage marks a total failure to grasp the conservation of substance, weight and volume, even during very slight deformations in shape. In the questionnaires from which the following extracts have been taken, we have been careful to ask separate questions about weight and substance, to make certain that the first of these qualities is not, in fact, quantified before the second:

LOU (4; 6) constructs a ball similar to the model he is shown: 'Is there as much clay in the second ball as in the first?— *Yes.*—Are they as heavy as each other?— *Yes.*—And as big?— *Yes.*' The two balls are flattened, the first slightly and the second more heavily, to produce one thick and one thinner but wider disc. 'Are they still the same?— *No. That one* (the thicker disc) *is heavier.*— Why?— *Because it's got more clay.*—Why?— *Because it's thicker.*' Similarly

[1] J. Piaget and A. Szeminska, *op. cit.*

5

when the first of the two discs is transformed into a coil while the second is restored to its original spherical shape, Lou thinks that the first is *'lighter.—* Why?—*Because it's got less clay.'*

BAT (4; 7) constructs a ball that is a precise copy of the model. One of the balls is changed into a short cylinder and the other into a long coil: 'Are they still the same?—*No, this one* (the second) *is bigger.*—Did the two round balls have as much clay as each other?—*Yes.*—And now?—*No.*—Does one of them have more clay than the others?—*Yes, the longer one.'*

MAR (5; 5) states that the two balls are *'as big as each other'* and also *'as heavy as each other'*. One of them is changed into a coil: 'Are they still as heavy as each other?—*No.*—Why?—*That one is heavier.*—Why?—*Because it's longer.*—Do they still have as much clay as each other?—*No.*—Why?— *There is more in that one.*—Why?—*Because it's long.'* The demonstrator draws out the coil into a long thread and the ball into a short coil: 'And now are they still as heavy as each other?—*No. that one* (the short coil) *is heavier because it's thicker.*—Is there still as much clay in both?—*No, there is more in that one* (the coil) *because it's thicker.'*

CHEV (6; 6). One of the balls is changed into a coil and the other into a thick disc: 'Are they still the same?—*No. That one* (the disc) *is heavier.*— Why?—*It is a tiny bit bigger.*—Did they have as much clay as each other?— *Yes.*—And now?—*No.*—Where is there more?—*There* (the disc).—Why?— *Because it's bigger.*—What do you mean by bigger?—*It is bigger because it is a little heavier than that one.*—But is there still as much clay in both?—*No, that one* (the coil) *has a little less.'* The two balls are restored to their original shape and Chev now says that they are quite the same. Next they are changed into two discs, one of which is thicker and the other of greater diameter: *'That one* (the thick disc) *is heavier than the other because it has more clay.'*

COP (6; 0). The two balls are changed into a disc and a short cylinder: 'Is there still as much clay in both?—*No, that one* (the disc) *has more.*—Why?— *Because it's bigger all round* (points to the thickness at the circumference).— But where did the clay of that one (the coil) go?—...—Isn't it the same as before?—*No.'*

JUN (7; 3). One of the balls is changed into a coil; the other is left as it was: 'Are they still as heavy as each other?—*No.*—Why?—*Because that one* (the ball) *is bigger.*—Do they have as much clay as each other?—*No, that one* (the ball) *has more.*—But why does that one (the coil) have less?—...—Did they have as much clay as each other before?—*Yes.*—So where did the clay from this one (the coil) go?—*A little of it has dropped on to the table.*—Is that so?—*No.*—So do they still have as much clay as each other?—*No.*—Which one has less?—*That one* (the coil).—Why? *Because some of it fell down.*— Where did it fall?—...'

ROG (7; 3). One of the balls is changed into a disc and the other into a cylinder: *'This one* (the cylinder) *is heavier than the other because it's thicker.*— Why does that make it heavier?—*Because there is more clay in it.*—Which one has more clay?—*That one* (the cylinder).—But before you told me that the two balls had as much clay as each other?—*Yes, I did say so, but now there is more in that one* (the cylinder) *than in this one* (the disc), *because it's thicker.*—But you did say that the balls had as much clay as each other?—

Yes.—Are they still as heavy as each other?—*No, this one* (the cylinder) *is heavier because there is more clay in it.*—Tell me, a moment ago there were two balls. Did they have as much clay as each other?—*Yes.*—Do they still have as much clay as each other?—*No, this one* (the cylinder) *has more because it's thicker.*—But where has the clay gone?—*You flattened it out, so now there is less.'*

FIL (7; 2). The two balls are changed into two cups, one thick and the other thinner and of greater diameter: 'Look at what I am doing. Are they still as heavy as each other?—*No, that one* (the thin cup) *is heavier.*—Why?—*Because it has edges.'* The cups are transformed into two discs, one flat and thin the other thicker and of smaller diameter: 'And now?—*That one* (the flat disc) *is heavier because it has been flattened out.*—Why does flattening make it heavier?—*Because there is more clay in it.*—Is there more clay in this one (the large thin disc) than in that one (the thick disc)?—*Yes, because there isn't very much there.*—But before they were the same?—*Yes, but now there is more in this one* (the thin disc).' The thick disc is changed into a cube: 'And that one?—*Ah, now there is more over here* (the cube) *because there is more clay in the middle of it.*—But before they were the same, so why should one have more now?—*It's got bigger.'*

PIE (7; 1). 'Look at these two balls, do they contain as much clay as each other?—*Yes.*—Watch.' One ball is changed into a coil: '*The sausage has more clay.*—And if I roll it in my palm and turn it back into a ball?—*I think it will be the same again.'* The coil is changed back into a ball, and the other ball is flattened into a disc: 'Do they still have as much clay as each other?—*No, the ball has more clay.'*

The child's grasp of the conservation of substance can also be determined by his reactions to divisions of the clay ball.

CAR (6; 0). The equivalence of the two balls having been granted, the demonstrator cuts one of them into seven small pieces, which are placed on one pan of a balance to demonstrate that they weigh as much as the undivided ball, placed on the other pan: 'Do they still contain as much clay as each other?—*No, the little bits have more.*—And if we put all the little bits together?—*That would make a large ball of clay, but it will be heavier* (than the fragments) *because it's a ball again.*—Why is that heavier?—*The small bits are lighter, but there is more clay in the pan.'*

LUC (6; 6). Same experiment: '*The small bits are lighter.*—And do they contain as much clay as each other?—*No, the big ball has more.'*

We see that these children are convinced that the quantity of matter increases or decreases upon all changes in shape. Moreover, they seem to lack any criterion for deciding in favour of increases rather than decreases: the choice varies from one child to the next and may even vary in one and the same subject from moment to moment. Thus while most of them (Jun, etc.) believed that the ball contains more clay than the coil, Pie *et al.* took the opposite view; the first group justified their opinion by claiming that the ball was 'bigger' (Jun) and the second by claiming that the coil was 'longer' (Mar). Similarly Rog thought that the cylinder had more clay in it than the flat disc 'because it's thicker', while Chev asserted that the thinner of two discs had more clay than the other 'because it's bigger'. In short, the child's answer depends on whether he focuses his attention on the difference in thickness, length or diameter, etc.

As for the justification of non-conservation as such, it constitutes no problem at all for the child: he thinks it self-evident that the quantity of matter must vary whenever the shape of an object is changed. If the demonstrator tries to draw his attention to the odd consequences of this interpretation, for instance by asking what has happened to the missing clay, the child will tell him variously: 'A little of it has dropped on to the table' (Jun), or 'You flattened it out, so now it's less' (Rog). However, if the demonstrator does not dwell on the subject and allows the child to come up with spontaneous explanations, he will generally find that changes in the quantity of matter are attributed to changes in weight and changes in weight to changes in shape. Exceptionally the opposite may happen, as when Car, for instance, assumed that increases in the quantity of matter lead to decreases in weight.

Must we therefore take it that the reason why these children fail to appreciate the conservation of matter is that they fail to grasp the conservation of weight or volume during the transformations we have just described? In fact, it is easy to show that the invariance of the quantity of matter is acquired well before that of weight and volume. Thus out of the 180 four- to six-year-olds whom we questioned in Geneva, Lausanne and Neuchâtel, 55 had not the least inkling of any kind of conservation; 67 accepted the conservation of matter but not that of weight or volume; 38 that of matter and weight but not that of volume; and some 20 the conservation of all three. To be sure, at the level we are discussing, i.e. before the idea of the conservation of matter is grasped, there is a relative failure to differentiate between quantity of matter, weight and volume, and this is precisely why the child at that stage argues in a circle; a given object is heavier than another because it contains more matter, and it contains more matter because it is heavier, etc. On the other hand, once he has mastered the idea of conservation in full (Stage IV), he has, as we shall see, also learned to derive the conservation of matter from that of weight and volume or *vice versa* and, moreover, in a coherent logical manner. But until that happens, the three types of conservation and all the quantifications and invariances they entail, remain distinct.

What precisely does the child at Stage I make of the quantity of matter and why does he not grasp that it is conserved? To begin with, we know that he treats matter as the most general of all qualities, and that he eventually quantifies it well before such of its special qualities as weight and volume. In other words, it is simply because the child at Stage I is not yet in possession of quantifying operations that he fails to grasp the conservation of matter. This is in full accord with our earlier studies,[1] which showed that when liquids or pearls are poured from one vessel into another of different size or shape, children at Stage I fail to grasp the conservation of quantity as such. However, in the case of the clay ball, the apparent changes in weight and volume of the original clay ball complicate the picture even further, with the result that the conservation of matter is not grasped until several months later than that of the liquids and pearls. But despite this delay, it is obvious that the reason for the non-conservation of physical matter must be the same as that of mathematical entities: the primacy of direct perception over intellectual operations, i.e. the failure to co-ordinate relations and the lack of operational reversibility. When subjects at Stage I try to justify an increase or decrease in the quantity of matter, they confine themselves to invoking just one of the relations in question ('it is longer', 'thicker', 'flatter', etc.) and completely ignore the rest; they do not realize that the differences cancel out once they are co-ordinated in a total system. Moreover they are not certain that the distorted ball of clay can be restored to its initial state or, when they are (e.g. Pie: 'I think it will be the same again') they view this process in purely empirical terms and not yet in terms of rational reversibility.

§3. *The second stage, Sub-stage IIA: intermediate reactions between non-conservation and conservation of substance*

Stage II sees the discovery of the conservation of substance, but not yet of weight and volume. Sub-stage IIA is characterized by reactions oscillating between those characteristic of the previous stage (Stage I) and those characteristic of Sub-stage IIB, at which the conservation of the quantity of substance is invoked straightaway as a logical necessity:

EXE (6; 0). Ball transformed into a coil: '*There is more clay in the ball* (than in the coil).' Ball divided into two: '*They have as much clay as each other.*' Ball divided into six: '*It's as much clay ... No, there is more over there* (the large ball) *than over here* (the set of small balls placed on the other pan of the scale) *... No, it's the same because nothing has been taken away.*'

JAQ (7; 0) thinks that there is more clay in the ball than in the coil or the disc. However, when one of the balls is cut into two and he is asked to compare

[1] J. Piaget and A. Szeminska, *op. cit.*, Chapters I and II.

the results with the original ball he says: '*I shall have to think. Ah, it's the same thing because if we turn these* (the two small balls) *into a* (single) *ball the two would* be the same.'

DAN (7; 0) also hesitates. When one of the balls is turned into a coil he says: '*It comes to the same because when you rolled it (*the coil) *out no clay was taken away.*' One of the small balls is cut up into five little balls: 'Is there as much clay in all these together as in the big one? — *No, there is less over here* (the five balls) *because it isn't so big.* — Can we turn them back into a large ball? — *Yes, then it'll be the same again.*'

ROUG (7; 6). One of the balls is changed into a coil: '*It's the same thing, you've used the same clay.* — Is there as much to eat in the one as in the other? — *Oh, no, that one* (the ball) *has more.* — Why? — *Because it's in a ball.*'

CHAR (10; 0, educationally backward). When one of the balls is changed into a coil and he is asked about the weight, he himself introduces the conservation of matter: '*When it is long like that, it loses a little weight. When it's in a ball the clay is all tight, but in a sausage it's what you would call more spread out.* — But what happens when it's tighter? — *It takes more clay.* — Does it really? — *Yes, there is less in the sausage.* — How do you know? — *This one is in a ball but that one is thin and long.* — Could we turn the sausage back into a ball? — *Yes.* — Of the same size? — *No, just a tiny bit smaller.*' The ball is divided into small balls: 'Is there as much clay as before? — *There is more now because it's in small bits.*' One of the two small balls is flattened into a disc: '*There is a little more now because it's spread out; no, it's the same thing.* — Can we turn it back into a ball? — *Yes.* — Will we have to add more clay? — *No, it will be* the same.' As for the small balls, Char sticks to his original opinion.

Like all intermediate reactions, these responses give us a far better insight into the underlying thought processes than more positive answers could possibly have done. The problem of conservation reflects a conflict between direct experience or perception and rational operations. So long as they rely on perception alone, children at this sub-stage argue like those of the previous stage: there is more clay in the ball because it is 'tight' (Char) or 'in a ball' (Roug), and there is less in the coil because it is more 'spread out', 'thin' or 'long', etc. (Char). But as soon as they stop relying on the appearances and begin to reflect on the transformations as such, they are forcibly led to the conservation of matter. The two operations that lead them there are: identification and reversibility.

The first is invoked more often than the second: 'Nothing has been taken away,' said Exe; 'no clay was taken away,' said Dan; 'you've used the same clay,' said Roug. Why then is this simple idea, i.e. that nothing has been taken away or added, not invoked earlier and why does it not lead the child directly to the conservation of matter? After all, children at Stage I know just as well as children at Stage II that the amount of clay in the original ball has not been increased or diminished during the successive deformations, so that when they say that 'some dropped on to the table' (Jun), they are merely trying to oblige the examiner with some sort of answer that will get them off the hook. But in that case why is identification not invoked until Stage II?

The reason is that identification as such is not enough to ensure conservation: identification can be applied to the data of perception only with the help of intellectual operations. In purely perceptive terms, the coil is not identical with the ball: it is less compact, thinner, etc., and hence suggests the idea of a decrease in substance. Before this idea can be dispelled the data must be elaborated with the help of a system of operations of which identity can only be the result and not the source.

It is at this point that reversibility enters the scene. Jaq showed this quite plainly when he justified his identification by an appeal to the inverse operation of division: 'It's the same thing because if we turn these (the two small balls) into a (single) ball the two would be the same.' Similarly we had only to ask Dan, who also did not believe in conservation, if the process of division could be reversed, to be told: 'Yes, then it will be the same again.' Char finished up with much the same reaction.

Now, whoever says reversibility also says direct and inverse operations, i.e. operational thinking. This is proved by the fact that a simple empirical return to the original shape does not suffice to ensure conservation, precisely because it does not yet constitute true reversibility. Thus children at Stage I sometimes grant that it is possible to return to the starting point (Pie, for example, thought that if the coil were re-shaped into a ball 'it will be the same again'), but they do not yet deduce the concept of conservation. What then is the precise difference between these two reactions? Obviously it is the difference between operational and non-operational thought. A simple empirical return to the initial state strikes the child as a possibility but not as a logical necessity ('I think,' said Pie), because, as far as he is concerned, what is involved is a purely intuitive succession of physical states characterized by their perceptible qualities alone. The child thus focuses his attention on the physical state as such and not on the transformations: hence even if he should grant that it is possible to return to state A from state B, this return in no way ensures the conservation in B of the quantitative properties of A. Instead of applying operational reversibility he relies on an intuitive process and thinks of a succession of distinct states. With true reversibility, by contrast, which appears during Stage II, the return to the starting point strikes the child as a logical necessity and no longer as an empirical possibility: he has grasped that the very operations responsible for the transformations are

11

reversible. No matter whether he is examining deformations or divisions the child at this sub-stage has begun to realize that every act of rolling out, flattening, elongating, cutting, etc. can be reversed by an opposite act that offsets the effects of the first. True reversibility is therefore the discovery of the inverse operation *qua* operation, and that is why the mental approach accompanying the transition from intuition to operationality leads quite automatically to a grasp of conservation. This is what we shall be seeing more clearly in our discussion of reactions at Sub-stage IIB: the only difference between it and Sub-stage IIA is that in IIA conservation and operational reversibility are still confined to small-scale deformations. This is because the operation involved (the reversible action) has not yet been divorced from intuition, i.e. from irreversible actions and from perception. Moreover, in small-scale transformations, the child's mind can surmount the perceptible appearances thanks to a grasp of the operations, but as soon as the deformations go beyond a certain limit, direct intuition comes to prevail over operational intelligence and conservation is again called into question. During Sub-stage IIB, by contrast, the operational mechanism finally parts company with perceptive intuition and the conservation of matter — unlike that of weight or volume — is acknowledged under all circumstances.

§4. Sub-stage IIB: the conservation of substance

Subjects at the stage we are about to discuss grant the conservation of substance in all circumstances, but refuse to grant the conservation of weight.

Here, to begin with, are some examples of a complete grasp of the conservation of substance in the case of simple changes in shape:

FRA (6; 6). One ball is changed into a coil: 'Are they still as heavy as each other?—*No. That one* (the ball) *is heavier.—*Why?*—Because it is bigger.—*Is there still as much clay in both?—*Yes.—*Why?*—Because there is the same clay as before.'

RAC (7; 6). 'Is there as much clay as before?—*Yes, because none has been taken away, so it* (the coil) *must be as big* (as the ball).'

APO (8; 2). One of the two balls is changed into a disc: '*It's the same thing. If we* (re-) *made it into a ball it would use up as much clay.*'

BER (9; 0). '*There is the same clay. It's still the same as the ball. You have just changed its shape.*'

NO (9; 0). '*It's the same thing as before. When it's drawn out* (into a coil) *or when it* (its shape) *is changed, it's still the same* (quantity of matter).—Why? —*It's longer, but it's thinner: it's still the same.'*

12

EV (9; 0). 'Is there as much clay as before?—*It certainly is.*—Why?—... —
But how do you know that it certainly is?—*It's obvious.'*

FOE (9; 6). *'First that one was round, and now it's been made long, but it has
just as much clay; you didn't take any away.*—Can we turn it back into a ball?—
Certainly we can, there isn't more clay in it, it's the same thing (laughs). *There
isn't any over at all; even if it is long now it comes to the same.'*

BUR (9; 11). 'Do they have as much clay as each other or does this one (the
coil) have more?—*Look here, rolling it out didn't make it lose any clay. It's
all the same. It's just as if it were* (still) *a ball.*—And if we turn it back into a
ball? Would it get smaller or bigger?—*It would be the same size, none of the
clay is missing.'*

RUG (10; 6). *'They are both the same, because it's the same amount, even
though one is longer now.*—Why is it the same amount?' He looks attentively
at the coil which is still being extended. *'I am looking to see if it's still the same
when it's all rolled out. Yes, it is. I guessed right, because you can turn it back
into the same ball.'*

GIV (11; 0). *'There is always the same clay, so there can't be more or less.'*

ROS (12; 0). *'It's the same because nothing was removed and none was added.'*

Here now are some examples of the grasp of conservation during
divisions of the ball of clay:

VA (8; 0). The ball is divided into five or six smaller ones: *'It's the same
amount but less big.*—Why?—*Because if they were stuck together again
they'd get flattened out and smaller, but it's still the same amount.'* (Clearly, a
grasp of the conservation of matter does not necessarily entail a grasp of
the conservation of volume.)

GAI (8; 0). The ball is divided into five small balls, but only after the child
has noted how much water (volume) the original ball displaces in a measuring
jar: 'Will these take up the same room in the water?—*They will fill up the
bottom just as much, but the small pieces will take up less room than the ball in
a lump.*—And the clay, is there as much, or is there more or less?—*It's the
same, only here it's in small pieces.*—How do you know that there is as much
clay as before?—*Because if we turn it back into a ball we shall see it's the
same.'*

BRU (9; 9). Disc and five little balls: 'Do they have as much clay as each
other?—*They certainly do. It's as if it were one ball, one single piece.'*

CHE (9; 2). *'It's the same. It's all the clay from the ball, but cut up.'*

GOL (10; 6). Large ball and seven small balls: *'It's the same and it's got the
same number on either side.*—The same number of what?—*Of bits, one
could make a single ball out of them by pressing them hard together: that
would make the same thing.*—How can we tell?—*Because it's the same clay.'*

We see how clear all these reactions are: the conservation of
matter is affirmed by all subjects as if it were inconceivable that it
should be otherwise. Thus while subjects at Sub-stage IIA give
different responses from one deformation of the ball to the next,
or only discover the conservation of matter in the course of the
test, those at Sub-stage IIB have an immediate, apodictic grasp

of the invariance : 'Certainly we can (turn the coil back into a ball),' said Foe, laughing at the thought that we could ever doubt it; 'Look here!' said Bur, as if he were giving us a lesson; 'it certainly is,' said Ev, though he could adduce no proof unless it was his own certainty.

This enables us to rule out a possible objection to our interpretation of Sub-stage IIA reactions, namely that, instead of being led to conservation by their grasp of operational reversibility, our subjects might, in fact, have arrived there because we kept restoring the clay to its original shape in the course of the test. But if this interpretation might have been true of subjects who were on the point of grasping the idea of invariance, it cannot possibly account for that sense of logical necessity so characteristic of subjects at Sub-stage IIB. The explanation must therefore be sought elsewhere, namely in the grasp of operations and the discovery that they are reversible. (See §3.)

Let us therefore look more closely at the true nature of the child's grasp of the conservation of matter or of substance. Here, his intellectual progress faces us with a paradoxical situation : on the one hand, the discovery of the conservation of substance precedes that of the conservation of weight and of volume; on the other hand, the conservation of weight and volume is eventually elaborated with the help of mechanisms that are identical to those we have just described (the same words are used to justify all three types of conservation). How, then, does the child view a substance that is apparently conserved well in advance of its attributes? We shall try to show that he views it as an undifferentiated and global quality, one that completes on the conceptual plane the global qualities of the sensori-motor object, and also that the conservation of this substance represents the simplest possible quantification of qualities, which sets it off sharply from the measurement of such differentiated, and consequently more complex, qualities as weight and volume.

Let us first recall that the process of quantification characteristic of what we have called the conservation of substance shows no formal distinction from the construction we have previously described as the conservation of continuous quantities.[1] Thus, in the case of the decantation of a fixed amount of liquid or a fixed number of pearls from one vessel into another of different shape, we saw that the child's construction of the conservation of quantity involved three successive and clearly distinct stages : that of non-conservation, during which the perceptive relations of length, breadth, etc. are not co-ordinated and the notion of total quantity has no meaning; an intermediate stage during which the

[1] J. Piaget and A. Szeminska, *op. cit.*, Chapter I.

child arrives at conservation in simple cases but relies on perceptive intuition for more complex situations; and finally the stage of complete conservation which appears at about the age of six or seven years. Now, the conservation of substance which is the subject of this chapter involves precisely the same stages, though, for reasons we have explained, with some delay.

If the conservation of substance is indeed the most elementary form of the conservation of continuous quantities, it follows that it serves the child as a general schema of quantification, i.e. as the simplest and most undifferentiated physical *quantum*. 'As much clay', 'the same amount', 'it's still the same', 'the same lump', 'the same piece', even 'the same number': it is in these terms that our subjects refer to the general conservation of quantity. Now, this raises a problem that did not appear during our study of the child's conception of number, namely to what precise qualities the child applies this schema of quantification and more especially what qualitative meaning he attaches to a substance that can be quantified well in advance of weight and volume. In short, the special problem to be examined in this book is the quantification of physical qualities as such.

Now the reason why substance is quantified before its attributes (weight and volume) is, first of all, that it provides young subjects with an undifferentiated quality, with a springboard from which they can advance to the special qualities that become quantified in the very course of their differentiation. The invariance of matter thus constitutes, in the second place, an extension of the invariance of objects but with one essential difference. In the case of objects the problem is to establish that they remain constant despite apparent changes in shape or size resulting from changes in the child's perspective, and to establish that the child needs no more than a purely intuitive form of logic, or even enough practical intelligence, to co-ordinate his perceptions into an appropriate spatial group of displacements. On the other hand, when the perspective remains unchanged, but the object (e.g. a ball of clay) is subjected to real deformations, then the problem of conservation is no longer to establish the invariance in shape or size, but the invariance of the support of what have become variable properties. In other words, the substantial quality has ceased to be particular and perceptible (length, breadth, weight, colour, etc.) and has become the quality of the permanent support of its variable attributes. Now, we believe that the conservation of this substantial support must necessarily lead to a quantification, or indeed imply it from the very outset. In fact, whenever an object changes its shape and dimensions, the permanence of its substance can be grasped only through the neutralization of these differences: now, this process demands the division of the clay into homogeneous parts or units (see below), which explains why substance becomes a *quantum* just as soon as the substantial quality is considered invariable. Moreover, as the particular qualities of the deformed object (e.g. its weight and later its volume) become quantified in turn, so the substance ceases to be an undifferentiated support, merges with these quantitative invariants, and hence loses its fictive and global character.

Briefly, the conservation of substance marks both the beginning of the quantification of qualities and the completion of the construction of objects. Substance is therefore on the one hand a kind of formal regulator, the significance of which will become clearer as and when the quantitative schema can be applied to differentiated or particular qualities, i.e. to weight and to volume, and to their gradual syntheses (atomism and relative density). On the other hand, while these qualities are unquantified, substance remains an undifferentiated quality providing the content of this general *quantum*.

This paradoxical situation highlights the problem posed by conservation, namely that the first principle constructed by the child explicitly involves the permanence of a *quantum* that is still undifferentiated and hence does not correspond to any sensible quality. Being schematic and global, this first principle cannot possibly be constructed without intellectual activity, i.e. without going beyond direct experience. Certainly, experience can suggest it, and, far from contradicting it, can derive a great deal of illumination from it. But the real basis of the construction of the first principle has to be sought in the workings of the intelligence, as witness our Stage IIB responses. Admittedly, subjects at this stage continue to justify the conservation of matter indifferently by identification and reversibility; but their identifications are all based on the idea of an operation. 'First that one (the ball) was round,' said Foe, 'and now it's long, but it has just as much clay; you didn't take any away.' 'Rolling it out didn't make it lose any clay,' said Bur. 'There is the same clay. You have only changed its shape,' said Ber. 'When it's drawn out or when it is changed, it's still the same,' said No most expressively; 'It's all the clay from the ball, but cut up,' said Che, etc. In other words, the identity has become associated with the operations themselves, thus losing its static character and introducing reversibility by implication. Some of the subjects, indeed, bring this aspect out explicitly: 'It's the same thing. If we (re-)made it into a ball,' said Apo, 'it (the coil) would use up as much clay.' 'It's the same amount (of clay) because you can turn it back into the same ball,' said Rug. Gai explained that 'if we turn it back into a ball we shall see it's the same'. Finally, Gol added that 'we could make a single ball out of them by pressing them hard together'. In short, every one of these subjects made use of deduction, and not of deduction based merely on simple identification but also, and above all, on the composition and inversion of the constructive operations: all realized that something was being conserved, though they did not yet know precisely what unless it was matter as such, i.e. an undifferentiated quality.

But in what way – and here we return to a point touched upon in the previous paragraph – is such operational reversibility the expression of logical deduction rather than the empirical finding that it is possible to restore the clay to its initial state? Now, an operation is neither a physical transformation nor some vague psychological act: it is a reversible action in the precise sense that it engenders relations (or classes) whose converse (or exclusion of these classes) is produced by the inverse action. This explains why a simple empirical return to the starting point is not enough to lead to conservation; it remains a mere possibility, and only becomes a necessity once the subject has come to appreciate that the differences due to the transformation of the object can be reversed and so cancelled out. Now, when the child says 'make it into a ball' (Apo), 'make it long' (Foe), 'roll it out' (Bur), 'change its shape' (Ber), 'draw it out' (No), etc., to describe the changes in shape and 'cut it into small pieces' (Hem), or 'stick them together' (Cla), etc., to describe the division of the ball or the opposite action, the operations he describes are real intellectual operations because they engender spatial relations or relations between the whole and its parts such that any difference can be cancelled out by the inverse operation.

Indeed, it is precisely because perceptive relations have made way for operational relations that the latter can be co-ordinated. In other words, the global actions of drawing out, flattening, cutting, etc., do not become reversible operations until they can be expressed, no longer merely as simple and discontinuous relations, but as complementary relations that can either be added together or multiplied among themselves. In fact, the main difference between children at Stage I and those at Sub-stage IIA is that the former consider the elongated shape of the coil (the relationship 'longer') or the compact shape of the ball (the relationship 'bigger' or 'thicker') as absolute and isolated qualities, whereas the latter realize immediately that though the coil is longer than the ball, it is also thinner, i.e. that the two relations must be multiplied. Similarly with the divisions of the ball: the partition of the whole into parts suggests the idea of material decreases to children at Stage I, simply because the pieces become smaller. Subjects at Stage II, by contrast, mentally add these parts into a whole, whose elements are the more numerous the smaller they are, so that these two relations once again cancel each other out.

This operational co-ordination of logical relations leads them directly to mathematical quantification. What, in fact, is the assertion that the transformation of the clay ball produces 'the same size', 'the same number' or 'the same quantity' if not the realization (at least implicitly) that the differences cancel out, that 'longer' × 'thinner' = 'the same quantity'? This is precisely what No stated in the most explicit terms when he said: 'It's longer but it's thinner; it's still the same.' Now this assertion which, at first sight, appears as the conclusion of a simple logical multiplication of two relations (increase in length × decrease in diameter) takes us far beyond qualitative operations. The point is worth stressing, as this problem will recur throughout this book.

It should be remembered that all logic presupposes quantification but of a simple type which, following Kant, we may call 'intensive' because it bears exclusively on the relationship between the whole and its parts. Thus from the fact that all Genevans (A) are Swiss (B) but some Swiss are non-Genevans ($\bar{A}$) we can deduce that $A < B$ and that $\bar{A} < B$, but we can say nothing at all about the quantitative relationship between A and $\bar{A}$. (We can have $A \gtrless \bar{A}$ or $A = \bar{A}$.) Similarly in a series of asymmetrical relations, if x differs from y (i.e. $x \xrightarrow{a} y$) and if y differs from z (i.e. $y \xrightarrow{a'} z$) we know that the difference between x and z (i.e. $x \xrightarrow{b} z$) is greater than that between x and y ($\xrightarrow{a}$) or that between y and z ($\xrightarrow{a'}$), but we know nothing at all about the relationship between $\xrightarrow{a}$ and $\xrightarrow{a'}$ which may just as well be $\xrightarrow{a} \gtrless \xrightarrow{a'}$ as $\xrightarrow{a} = \xrightarrow{a'}$.

'Extensive' quantification, by contrast, appears as soon as we compare the parts to one another quantitatively, e.g. A to $\bar{A}$, or $\xrightarrow{a}$ to $\xrightarrow{a'}$. Lastly, we speak of 'metric' quantifications to describe a third type of quantity which is a special case of the second. It appears when, the parts (or the differences) having been equalized, it is possible to introduce the idea of unity: i.e. if $A = \bar{A}$, then $B = 2A$; and if $\xrightarrow{a} = \xrightarrow{a'}$, then $\xrightarrow{b'} = \xrightarrow{2a}$.

Let us now try to see what logical schemata are introduced by the operations that lead the child to the conservation of matter, and what quantities intervene in them. Take a lump of clay C_1, and remove a mass B'_1 which can be divided into several parts: you are then left with $C_1 - B'_1 = B_1$. Having detached B'_1 from one of the extremities of B_1, let us now change its shape and return it to another, different part of B_1. We now have a new lump, C_2, such that $C_2 = B_1 + B'_2$, where B_1 is the portion that was left intact, and B'_2 the part that was added to form the new arrangement. In that case, how can we prove that $C_1 = C_2$? By four methods and four only: (1) identification of the elements (classes or parts); (2) equalization of units; (3) identification of the spatial relationships; and (4) equalization of the differences.

In respect of (1), $B'_1 = B'_2$ can be established directly by qualitative identification of the constituent elements. If, for example, B'_1 is made up of the segments A_1, A'_1, etc., which can also be recognized in B'_2 or whose displacement the subject can follow in his mind, we have the logical equalities (identities) $B'_1 = B'_2$ and $B_1 = B_1$; whence $C_1 = C_2$. (2) Next, let us assume that the segments A_1; A'_1; etc., are equal to each other: in that case we can count them off. If $B'_1 = nA$ and $B'_2 = nA$ as well, then $B'_1 = B'_2$ once again, and if $B_1 = xA$ then $C_1 = n + xA$ and $C_2 = n + xA$ so that $C_1 = C_2$. This operation obviously goes beyond qualitative logic, since in order to equalize $A_1 = A'_1 =$ etc., the child must disregard the differential qualities by which he identified these elements in the first method.

(3) We can proceed also by the identification of the spatial relationships. Let C_1 be of any simple shape, and of length $\xrightarrow{c_1}$ and height $\downarrow b_1$; and let B_1 have the same height $\downarrow b_1$ and the length $\xrightarrow{b_1}$; and let B'_1 also have the height $\downarrow b_1$ but the length $\xrightarrow{b'_1}$. If we simply shift B'_1 beneath B_1 and convert its length $\xrightarrow{b'_1}$ into its height $\downarrow b'_2$, we have $C_1 = \downarrow b_1 \xrightarrow{b_1 + b'_1}$ and $C_2 = \xrightarrow{b_1} \downarrow b_1 + b'_2$. It is then immediately obvious that C_2 has gained in height, with respect to C_1, what it has lost in length, since $\xrightarrow{b'_1} = \downarrow b'_2$ (identity of relations). Hence $C_1 = C_2$.

(4) Now suppose that C_1 has a more complex shape and that it is no longer possible to identify $\xrightarrow{b'_1}$ with $\downarrow b'_2$. In that case, we can think of the differences in the spatial characteristics of the object as lending themselves to composition into units or ratios. Thus let C_1 be a cylinder and let it be pulled out into C_2, the respective diameters being $d_1 < d_2$ and the respective altitudes $h_1 > h_2$. Hence $d_1 \times h_1 = d_2 \times h_2$, so that we have the inverse ratio $\dfrac{d_1}{d_2} = \dfrac{h_2}{h_1}$.

To sum up: while methods 1 and 3 lead to intensive quantification in which every part is simply compared to the whole or to itself (identity), methods 2 and 4 involve extensive quantification (by the equalization or comparison of the parts), method 2 being metric while method 4 does not necessarily have to be so.

It should also be noted that, from the logical point of view, methods 2 and 4 are based on the same operations but applied to objects and to spatial relations respectively; the elements themselves (method 2) or their dimensions (method 4) are reduced to a system of real or virtual units.

Moreover, as we have tried to show elsewhere, first from a

logistic[1] and again from a psychological point of view,[2] every system of units results from the operational fusion of a grouping of classes and a grouping of asymmetrical relations. Hence methods 2 and 4 may be considered inseparable and as both resulting from a fusion of methods 1 and 3.

These conclusions are fully borne out by the reactions of our present subjects. On the one hand, the arguments they use when presented with the divided clay ball testify to the use of methods 1 and 2. Thus when Bru said, of the small balls: 'It's as if it were one ball', or when Che contended that 'it's all the clay from the ball, but cut up', they were treating fragments either as elements whose qualitative identity is preserved, no matter whether they are combined or separated, or else as units whose sum is equal to the whole ball. 'It's got the same number on either side,' Gol even said, when comparing the segments to the original ball.

By contrast when, as in the case of deformations without division, the child relies on the co-ordination of the relations and on the equalization of the differences, it is clear that he is employing methods 3 and 4. Thus when No and some others said, 'It's longer, but it's thinner; it's still the same', they meant either that the relations involved cancel out by qualitative identification (within a grouping of reversible operations), or else that their equalization calls for their reduction to a common standard (i.e. to units) or to proportionality. Finally, it is clear that these diverse operational procedures, some corresponding to the division of matter and others to the correlation of differences, are complementary and psychologically interdependent. This explains why the grasp of conservation with the help of a logical grouping (intensive quantification) goes hand in hand with a grasp of extensive quantification, and why methods 2 and 4 result from the combination of methods 1 and 3.

We may therefore take it that the reason why, once the conservation of substance has been constructed in all its generality, our subjects treat it as an *a priori* necessity and not as a simple empirical supposition, is that it results from the simultaneous grouping of logical operations and their mathematization: logical conservation, as it were, is extended directly and automatically into quantitative conservation. Now, if we return to our comparison between the substantial invariant, and that of the simple object of perception, the concrete meaning of these operations becomes obvious. The object of perception is an indivisible whole which retains its shape

[1] *Proceedings of the Société de Physique et d'Histoire naturelle de Genève*, 1941, 58, pp. 122–6.
[2] J. Piaget and A. Szeminska, *op. cit.*

and dimensions despite any apparent changes. During real deformations of the object, what is conserved, on the contrary, is no longer the perceptible whole but the sum of its parts considered as so many invariant objects. In short, while, before conservation, substance is no more than an undifferentiated quality serving as a support for the rest, as soon as it is treated as a constant and hence quantified, it *ipso facto* becomes a quality shared by all the small grouped objects constituting the whole. That the logical and quantitative composition leading to the conservation of matter presupposes this kind of partition into homogeneous units, and, what is more, an implicit or even an explicit adoption of atomism, will become quite clear after a further discussion of the relationship between substance, weight and volume.

CHAPTER TWO

The Conservation of Weight upon Deformations of a Clay Ball[1]

In the last chapter we tried to show that the grasp of the conservation of matter calls for a grasp of the reversibility of the transformations of the clay ball, i.e. for a co-ordination of the relations engendered by these operations, and also for an intensive and an extensive quantification. But why does this reversible and quantifying co-ordination not lead directly to the conservation of weight and volume?

As far as the conservation of weight is concerned, we shall see that its construction proceeds by the same stages as the conservation of matter, but instead of being completed by the age of eight, it is not reached until about the age of ten. The reason is that the conservation of weight calls for a quantification of more complex qualities, i.e. of qualities closely bound up with the subject's own activities. In fact, the child learns very early on to identify weight with the muscular effort needed to lift objects, and hence to connect weight with movement.

Let us recall that the stages passed in the construction of the conservation of weight are: (1) Stages IIA and B: conservation of substance but not of weight; (2) Stage IIIA: conservation of substance and intermediate reactions to the conservation of weight; and (3) Stage IIIB: conservation of substance and of weight but not of volume.

§1. *The second stage (Stages IIA and IIB):*
no conservation of weight

We shall deal separately with responses to deformations and divisions of the clay ball, beginning with two Stage IIA responses to deformations (intermediate reactions in respect of the conservation of substance with non-conservation of weight):

---

[1] In collaboration with Kiazim Osman.

VIS (5; 6). Two identical balls are produced: 'Are they as heavy as each other?—*Yes.*—Will the balance stay level?—*Yes.*' One of the balls is transformed into a coil: 'Are they still as heavy as each other?—*No, the second one* (the coil) *is heavier.*—Why?—*Because it's bigger.*—Why?—*Because it's longer.*—Has it as much clay as the other?—*Yes.*' The coil is changed into a long thread: 'Is it as heavy as before?—*That one* (the thread) *is heavier because it is bigger.*—Does it have as much clay?—*That one has more, because it's longer.*—I shall now make two threads (of the same length). Are they as heavy as each other?—*Yes.*—As much clay?—*Yes.*' The ends of one of the threads are joined together to make a ring: 'And now?—*The first one was heavier.*—Why?—*Because it was longer.*—Did it have as much clay as the other or not?—*The same.*—So is it as heavy?—*No, the first was heavier.*—Why?—*Because it* was bigger.'

The subject is shown two wads of cottonwool of the same dimensions: '*They're as heavy as each other.*' One of the wads is teased out: 'And like that?—*No, that one* (the loose wool) *is heavier.*—Why?—*Because it's bigger.*' Two small packets of tobacco of the same dimensions: 'And these?—*They are as heavy as each other.*' One of them is teased out, 'And that?—*That's heavier.*'

BON (6; 0). One of two balls is changed into a coil: 'Does the weight remain the same or not?—*That one* (the coil) *gets heavier because it's bigger.*—Did they have as much clay as each other before?—*Yes.*—And now?—*The round one has most.*—And which one did you say was heavier?—*The round one, the other has become lighter.*—Why?—*Because it's thinner.*' The ball is changed into a coil and the coil into a ring: 'And now?—*The first one* (the coil) *is heavier because it's longer.*—Do they have as much clay as each other?—*Yes, because none of the clay has been taken away.*—And are they as heavy as each other?—*The round one* (ring) *is heavier.*—Why?—*Because it is rounder.*—And if we make two balls out of them again?—*They'll be as heavy as each other, because both of them will be the same.*'

Two wads of cottonwool: 'Are they as heavy as each other?—*Yes.*' One wad is compressed and the other teased out: 'And now?—*The first one is lighter because it's smaller, the other one is heavier because it's bigger.*—Can we make a wad the same weight as before out of the smaller one?—*Yes.*—And if I squash the other will it have the same weight?—*It will be less heavy.*'

Here now are some clear second stage reactions (Stage IIB; conservation of substance but not yet of weight):

OC (6; 0). Ball and coil: 'Did they have the same weight before?—*Yes.*—And now?—*The first one* (the ball) *is heavier.*—Why?—*The clay is harder.*' The coil is turned back into a ball: 'And now?—*The same weight.*' Long coil and ring: 'And now?—*The second is heavier.*' The ring is opened up to produce a coil as long as the first one: 'And now?—*They are as heavy as each other.*' The coil is tied into a knot: 'And now?—*This one is heavier. That bit* (the top of the knot) *weighs down on the rest* (!).' The two coils are changed back into balls: 'And now?—*They weigh the same.*' One of the balls is drawn out: 'And now?—*The ball is heavier.*—Do they have as much clay as each other?—*Yes: it's like the first ball.*—And do they have the same weight?—*The ball weighs more.*'

'Now look, if we take a piece of cheese and grate it will there be as much cheese as before?—*Yes.*—As heavy?—*No.*—If I grate this ball will there be as much clay?—*Yes, none will have been taken away.*—Will it weigh the same?—*No, the round ball is* heavier because it hasn't been grated.'

MIN (6; 6). One of the two balls is changed into a ring: '*The first one* (the ball) *is heavier.*—Why?—*Because you've made the second one thinner.*' The ball is changed into a disc: 'And now?—*It's the same.*—Why?—*They are both thinner now.*' Ball and coil: 'And now?—*The first one is heavier.*—Why?—*It's not so thin.*—Do they have as much clay as each other?—*Yes.*—So why is the second one heavier?—*Because it has been pulled out.*'

One of the identical wads of cottonwool is teased out: '*The tight one is heavier.*' Cheese: '*It's heavier when it's ungrated.*'

SUZ (6; 6) examines the two balls: '*Oh sure, they weigh the same.*—And if I pull this one out into a thread will they still weigh the same?—*We'll have to see.*' One of the two balls is pulled out: '*No, the ball is quite heavy but that one weighs a little more, you've pulled it out so it's bound to weigh more.*—Can we turn it back into a ball?—*Yes.*—Will it get bigger or smaller?—*I don't know; oh, it'll be the same because it was a ball before.*—Do the two have as much clay as each other now?—*Yes.*—And the same weight?—*No.*'

AND (7; 0). '*That one is lighter because it's a sausage.*—How so?—*Because it's thinner.*—Is there as much clay here as there?—*Yes.*—And if we make the sausage a little fatter?—*The ball will be a little heavier all the same: that one* (the coil) *is round but the ball is rounder.*'

PHIL (7; 0). '*The coil is a bit lighter because it's thinner. This one* (the ball) *has a little more weight because it's squashed together.*—But do they have as much clay as each other?—*They do.*'

MOR (7; 0). One of the balls is changed into a cylinder: 'Do they still weigh the same?—*No.*—Why not?—*The ball is heavier because it is big and round.*—Do they have as much clay as each other?—*Yes.*—Then why is the second one heavier?—*Because you've pulled it out.*'

Same reaction with the wads of cottonwool: '*The squashed one is heavier.*—Why?—*Because it's rounder and the one you've pulled apart is lighter.*'

GAI (8; 0). Ball and coil: '*The longer it is the less it weighs. When it's squashed together it's heavier.*—Why?—*It's thicker.*' The coil is turned back into a ball, and the second ball is changed into a disc: 'And now?—*The ball is heavier, you can feel that it is when you pick it up. You can see that this one* (the disc) *is lighter; when it's flat it's less heavy than when it's in a ball.*—But do they still have as much clay as each other?—*Of course they have.*'

ROU (9; 0). '*The disc weighs more because it comes to the edge of the pan. So it weighs more.*' Also: '*the ball is heavier than the sausage.*—Why?—*Because it's bigger.*—Do they have as much clay as each other?—*Yes, because you didn't take any off.*—Are they as heavy as each other?—*No, because this one* (the coil) *is smaller.*'

ADO (10; 2). The red ball is left unchanged and the blue one drawn out into a coil: 'Do they still weigh the same?—*No, the red one is heavier.*—Why?—*It hasn't been let out and the blue one has.*—What do you mean by 'let out'?—*It makes it less heavy.*—Is there still as much clay?—*Yes.*—But the weights are different?— *Yes.*' The red ball is drawn out into a short coil and the blue one flattened into a disc: 'Do they still have as much clay as each other?— *Yes, because you haven't changed the clay.*—And the same weight?— *No, the red one is heavier because it's tighter, the blue one is lighter because it's looser.*— If you take a handkerchief and fold it up does it still have the same weight?— *No, it gets heavier; when it's folded up it's tighter and so it's heavier.*— And if we put these two balls on a scale won't they have the same weight?— *The red one will go down.*' The two balls are drawn out into coils of equal length, a fact to which the subject's attention is specially drawn. Next the blue coil is closed into a ring: 'Do they have the same weight?— *No, the blue one is heavier 'cause it's tighter and round.*—And when we grate a piece of cheese, do we change its weight?— *The ungrated cheese will be heavier.*—Are you sure?— *Not very.*—When we turn a ball into a coil, are you absolutely certain that its weight will change?— *Oh, yes.*—You have no doubts at all?— *No, I haven't.*—But some boys tell me that the weight does not change.— *They're wrong; the weight can't stay the same because it's been pulled out.*— And on the balance?— *It will be heavier on this side.*'

MEL (10; 0). '*The ball is heavier.*—Why?— *It's in a ball and the other one* (the coil) *is thin.*—But why is it heavy when it's in a ball?— *Because it has more weight, and because that one* (the coil) *sticks out* (beyond the edge of the pan).—And what does that mean?— *It weighs a bit less.*—And if I re-arrange it on the pan (in a semicircle) so that it won't stick out?— *That would make it a bit heavier, but all the same not as heavy as the ball. When it's long like that, a bit of the weight gets taken away. It's more spread out, but when it's in a ball the clay is all squashed together.*—And if I turn the coil back into a ball?— *It would weigh less, or perhaps it might get the same weight back.*—Why?— *It used to be a ball, and we saw that it had the same weight.*'

The two balls are reconstructed and one is flattened into a disc: '*The disc is heavier because it's flat and the ball is round.*—Why is it heavier when it's flat?— *Because there's more of it to touch the plate.*' Coil and disc: 'And now?— *The disc is heavier because it's flat; that one* (the coil) *is long and a little bit of it sticks out.*' Two discs, one arranged horizontally, the other vertically: 'And now?— *This one* (the horizontal disc) *is heavier than the one standing up because that one* (the vertical disc) *doesn't touch the plate all that much.*'

At the end of the test, Mel himself decides to weigh the ball and the disc and looks very surprised to find that they balance. He moves the ball closer to the centre of the pan and flattens the disc out further. Finally he weighs the coil and the ball: '*It's always the same weight!*—Why?— *This one* (the coil) *is lengthwise; you would think it's heavier like that, but the ball has more weight because it is right in the middle. When it's in the middle it's heavier than when it's not.*' Not even the balance can persuade Mel that the weight has not changed!

GRA (10; 6) also asserts that the ball is heavier than the coil. After weighing both in his hands, he says: '*When it's drawn out it is lighter because it's not in a lump; I could feel that this one* (the ball) *is heavier.*'

MUL (10; 0). '*The ball is lighter because it takes up less room on the balance; the coil is heavier because it's longer and thinner.*'

In short, all these subjects seem convinced that changes in shape entail changes in weight, yet except for Vis and Bon (Stage IIA) all of them also assume the conservation of matter. Now why do they fail to conclude directly from the latter as to the conservation of the former or, more precisely, why do the operations whose reversible co-ordination has led them to the invariance of matter not lead them automatically to the invariance of weight? The reason is plainly that the quantification of the qualities inherent in weight comparisons presents quite other problems than the quantification of matter *qua* undifferentiated quality. The time-lag between the construction of the conservation of matter and that of weight thus highlights the general problem of the quantification of physical qualities.

For a clearer understanding of the new problems raised by the quantification of weight we can do no better than look at the way in which our subjects try to justify the non-conservation of that quality. We find that, in addition to the reasons invoked at Stage I in favour of the non-conservation of matter, they come up with a whole series of special arguments based on what one might call the initial egocentrism of the concept of weight.

To begin with, most of them believe that when the ball is converted into a coil it loses weight because it is less compact, 'pulled out' (Min, Gra, *et al.*), 'longer' (Gai, Mel, *et al.*) or 'thinner' (Bon, Min, And, Phil, Mel, *et al.*), whereas the ball is 'round' (Bon, And, Mel), 'tighter' (Phil, Mor, Ado and Mel), 'thicker' (Mor, Rou), or 'squashed' (Gai). Now these were the very arguments children at Stage I (and even at Stage IIA) used to justify the non-conservation of matter, and which subjects at Stage IIB have come to reject. Why then is it that the same arguments, indeed the same words, that have lost their force in the case of the conservation of matter should still be invoked with such assurance in the case of weight?

The reason becomes clear when we recall the conditions governing the quantification of relations, i.e. the equalization of differences which we have examined elsewhere[1] and also in Chapter 1, §4 of this book. Take a lump of clay of height $\uparrow b$ and length $\xrightarrow{a}$, the two symbols representing the qualitative relations differentiating these magnitudes from zero. Now suppose we draw this lump out into a longer and lower shape: we now have $\xrightarrow{a} + \xrightarrow{a'} = \xrightarrow{b}$ and $\uparrow b - \downarrow a' = \uparrow a$. Neglecting the third dimension (equal to the height) so as to simplify our formulae, we can then say that the child postulates the conservation of matter just as soon as he comes to appreciate that these two differences ($+ \xrightarrow{a'}$ and $\downarrow a'$) cancel out. In other words: $\uparrow b \xrightarrow{a} = \uparrow a \xrightarrow{b}$ because $\downarrow a' = \xrightarrow{a'}$.

[1] J. Piaget and A. Szeminska, *The Child's Conception of Number*, Routledge & Kegan Paul, 1952, Chapters I and XII.

In concrete terms, this is precisely what the subject tries to express when he says that the coil loses in height what it gains in length from the lump of clay. Now as we saw in Chapter I, this step involves a mathematical quantification (even if no figures intervene) as soon as the equalization of the two relations does not reduce to a simple permutation of the height with the length, i.e. as soon as the child realizes that the totality $\uparrow b \xrightarrow{a}$ can be expressed by means of a constant system of (direct or inverse) ratios or even by spatial units whose product remains the same regardless of the arrangement of the elements. But if the extensive quantification of the relations reduces to this very simple schema, we still want to know why it should be easier to equalize the differences in the case of matter than it is in the case of weight.

Now that question was answered quite explicitly by Gai, Mel, Gra and others. The reason why all of them realized that a ball transformed into a coil conserves its quantity of matter was that it is relatively simple to discover that mere displacements do not alter the nature of matter : the parts of the ball which have been pushed down (i.e. $\downarrow a'$) help to increase the length (i.e. $\xrightarrow{a'}$), and the child need merely grasp the fact that the differences between the coil and the ball cancel out to appreciate that the quantity of matter remains constant (substance being nothing but the undifferentiated quality serving as the content of this elementary form of quantification). By contrast, in the case of weight the child has to decide whether a lump of clay taken from the top of the ball will weigh more, as much, or less than when it is shifted to the end of the coil. Now, to subjective experience the weight of a given substance seems to vary inversely with its tactile surface, i.e. the lump seems the heavier the less it is spread across the palm. This is what Mel tried to explain when he argued that the ball was heavier because it pressed down on the pan with all its weight, while the coil protruded over the edge so that part of it weighed nothing. To him it was quite obvious that the weight of $\downarrow a'$, i.e. of the clay removed from the top and bottom of the ball, could not possibly be equal to the weight of $\xrightarrow{a'}$, i.e. of the clay added to the extremities of the coil, even though he readily acknowledged that the quantity of matter was conserved. When the whole coil was fitted into the pan immediately afterwards, Mel continued to argue in the same way : 'That would make it a bit heavier, but all the same not as heavy as the ball. When it's long like that, a bit of the weight gets taken away. It's more spread out, but when it's in a ball the clay is all squashed together.' Gra was more explicit still, for not only did he put forward the same argument but he stuck to it even after he had weighed the ball and coil in his hands : 'When it's drawn out it is lighter because it's not in a lump; I could feel that this one (the ball) is heavier.' Gai, too, put it all most clearly when he said: 'The longer it is the less it weighs. When it's squashed together it's heavier.' In short, what prevents these subjects from equalizing the apparent differences in weight, i.e. from granting that the ball conserves not only its substance but also its weight, is that, in terms of subjective muscular impressions, the coil has not, in fact, gained in length what it has lost in height. A flat object seems to press down more lightly on the hand than one concentrated in a single point, and our subjects take it for granted that the balance will bear them out. This explains why Mel felt so sure that the ends of the coil protruding over the edge of the pan had no weight, or why Gra thought that the ball must be heavier than the coil.

Now, oddly enough, those subjects who take the opposite view, i.e. who contend that the coil is heavier than the ball, because it is more 'stretched out', argue precisely in the same way though in the opposite sense. Thus when Suz said of the coil: 'You pulled it out so it weighs more', and when Mul said: 'The ball is lighter because it takes up less room on the balance; the coil is heavier because it's longer and thinner', they were simply trying to say that the more space the clay takes up on the pan the heavier it must be, and this because a lump of clay that touches the pan directly must weigh more than anything added to the top of that lump and hence more remote from the pan. In short, the child finds it impossible to proceed to the equalization of the differences and hence to a quantitative partition, and this because he cannot render the parts homogeneous. Thus he readily acknowledges that if a given lump of clay is removed from the ball or from the end of the coil it will contain the same quantity of matter but he believes that the weight changes and this precisely because he evaluates it by means of sense impressions.

These two types of evaluation, contrary though based on the same explanatory principle, also appear during the comparisons of the disc with the ball, and even more expressly so. Most of our subjects contend that the disc is lighter than the ball because, as Gai put it, 'You feel that it is when you pick it up. You can see that this one (the disc) is lighter; when it's flat, it's less heavy than when it's in a ball', or 'when it's squashed together it's heavier'. Ado, for his part, thought the ball was heavier because it was 'tighter', and the disc 'lighter because it was more "let out"' and that, in that form, though containing the same amount of matter, it was 'less heavy'. He also maintained that a folded handkerchief gets 'tighter and so it's heavier', which, indeed, agrees with the subjective impression. Hence his general conclusion: 'The weight can't stay the same because it's been pulled out.' To other subjects, by contrast, the disc is heavier because it occupies the entire surface of the hand or of the pan. Thus Rou declared that 'the disc weighs more because it comes to the edge of the pan'. This is what Mel stated even more explicitly when he said: 'The disc is heavier because it's flat ... there's more of it to touch the plate.' Hence also his extraordinary claim that a disc standing on edge weighs less than one lying down, because it 'doesn't touch the plate all that much'. Nothing could highlight the problem posed by the quantification of weight better than Mel's refusal to equalize the differences in height and width even in the case of a mere change in the position of an object. In other words, he refused to grant the equality ($\uparrow a \xrightarrow{b} = \uparrow b \xrightarrow{a}$) because he believed that a' changes weight when it is applied vertically ($\uparrow a'$) rather than horizontally ($\xrightarrow{a'}$).

The same contradictory reactions, but based on a purely intuitive evaluation of weight, also resulted from comparisons of the two coils, one straight and the other closed into a ring. According to some subjects (e.g. Bon and Ado) the ring was heavier because it was round. According to others, however, the coil was heavier because it was longer. This is in full accord with our other findings, though Oc's comment is worth special mention because it reveals the mechanism on which all these explanations are based. After he had granted that two coils of equal shape had the same weight, the first was tied into a knot and he at once exclaimed that the knotted one was heavier because 'that bit (the knot) weighs down on the rest'. There is no better way of expressing weight in terms of the subjective impressions: egocentrism stands in the way of all extensive, and even intensive quantifications, for it prevents the equalization of the parts and their logical addition into a constant whole.

We can now see the true reason why these children attribute changes in weight to changes in shape. To them weight is a force that is not proportional to the mass, but can be considered a kind of active pressure depending both on its point of application and also on the shape of the body exerting it. This is why, to Oc, the ball was heavier than the coil, although it contained the same quantity of matter: its clay was 'harder', i.e. more compact and, as it were, more synergetic. Similarly Mor obviously considered weight a kind of pressure that decreases with dispersion of the substance: much as Mel found the ball heavier because it was 'squashed' together and the coil lighter because the clay in it was 'more spread out', so Mor contended that the ball was heavier because 'it is big and round'. He added that when the ball is transformed into a roll it becomes lighter 'because you pulled it out', as if greater adhesion of the parts ensured greater weight. This brings us to the wads of cottonwool and the tobacco. While some of our subjects simply considered the teased-out cottonwool or tobacco heavier because the wad had become more voluminous (Vis and Bon) – which is in keeping with the subjective impression that the bigger an object the heavier it is – others, and by far the greater number, thought that the smaller wad was the heavier because it was 'tighter' (Mor, Min, Oc, *et al.*). This was also Mor's attitude to the ball of clay: its weight increases with pressure and consequently with spatial concentration. In the same way, the cheese was generally thought heavier than the parings.

In short, all these reactions are convergent: weight to the child at this stage is not a physical constant independent of the shape of the object: different shapes are believed to exert different pressures on the weighing or evaluating subject, and the resulting impressions are projected on to the balance, which is thought to reflect the spatial contact between objects being weighed and the pans on which they are placed. In short, weight is not yet an objective relation: it is an activity, a function of muscular experience, whose manifestations are thought to vary with the effects they produce on the subject.

Nor should this attitude astonish us. After all, children at this stage have not yet advanced to an intensive, and *a fortiori* to an extensive, quantification of what to them is still a fluctuating quality, let alone to its conservation. The conservation of a quality, as we saw in Chapter 1, calls for the reversible co-ordination and for the quantification of the relations by which it is expressed. Now, if we are right in thinking that, upon becoming extensive, this quantification leads directly to the division of the whole into homogeneous parts, i.e. to the equalization of the differences distinguishing these parts, it is clear that this kind of operation cannot be applied to the case of weight while the child still believes that a part of one and the same whole (of the same substantial invariant) changes weight with changes in position, i.e. while he still questions the qualitative identity of the parts.

Let us now look at the corresponding responses to divisions of the balls or coils:

oc (6; 0). One of the two balls is divided into 9 small ones: 'Do they have the same weight?—*The big one weighs more.*—Why?—*Because it's larger, so it's heavier.*—Is there as much clay in the big ball as in all the little ones put together?—*Yes.*—So do they have the same weight?—*No, the big ball is heavier.*—Why?—*Because it's bigger.*'

min (6; 6). One of two identical coils is divided into seven pieces: 'If I put these on one side of the balance and all the rest on the other side will they have the same weight?—*No, this one* (the undivided coil) *will be heavier.*—

Why?—*Because it hasn't been cut up.*—And if I put these little coils together again, will they be the same as before?—*Yes, they will.*'

OSR (7; 10). One large ball and seven small balls: '*That one* (the large ball) *is heavier because it's all in a piece and these* (the seven small balls) *will be lighter because they are not in a piece.*'

ROL (7; 11). Same question: '*This one is lighter because it's smaller.*'

BUD (7; 6), by contrast, thinks that five segments weigh more than the undivided ball '*because there are more pieces*'.

GAI (8; 0). One of the balls is divided into six pieces: '*They are not as heavy as the ball.*—Why?—*When it's all in small pieces it's lighter than when it's in one piece; the big ball is heavier.*—Why?—*Because you can see that the small bits are thinner.*' But he thinks it is possible to make '*one big ball the same as before*' out of the segments.

DAL (9; 6). The pieces weigh less '*because they have been separated; they are thinner so they must be lighter. When they are all squashed together they are heavier.*—Why?—*They've got no weight left, they're too small and then they are at the edge of the pan.*'

ADO (10; 2). '*The red lot* (four segments) *is heavier because they're all spread out.*' The red segments are combined into a single ball and the blue ball is divided into four: '*The red one is heavier because the blues are more spread out.*' He thus contradicts himself. The red and the blue are divided into four segments each, with the blue segments almost touching one another: '*The blue lot is heavier because the bits are closer to each other.*'

MEL (10; 0). '*The small bits are heavier, because they're all over the pan, and that one* (the ball) *is right in the middle* (of the pan).—Can we make a ball out of them again?—*Yes, then they'll have the same weight because we'll have a ball, and then it will be right in the middle. The little bits take up more room and so they are heavier.*'

GOT (11; 0). '*The ball is heavier.*—Why?—*Because you can hold it better in your hand.*—But will they weigh the same on the balance?—*No, the bits are lighter because their weight is more spread out on the balance.*' The ball is changed into a disc which the subject is asked to compare with the fragments: '*The disc weighs less because it's larger and thinner.*—Why does it weigh less?—*Because the weight is more spread out.*' Pyramid and segments: '*And now?*—*The cornet weighs a little more because it is bigger.*'

These questions clearly elicit the same types of response as the previous ones. All these children realize perfectly well that the fragments contain as much substance as the undivided ball: 'It's the same, only here it's in small pieces,' Gai said referring to the substance (Chapter 1, §4), and 'It's all the clay from the ball, but cut up' (Che). Now the obvious fact that the sum of the parts is equal to the whole is neither invoked nor even recognized by these children when speaking of the weight. For most of them the sum of the fragments weighs less than the original ball, simply because the latter is 'bigger' (Oc), 'hasn't been cut up' (Min), or because the former are 'not in a piece' (Osr). Got produced a particularly

31

striking example of this attitude when he said that the big ball was heavier 'because you can hold it better in your hand', and that the 'bits are lighter because their weight is more spread out on the balance' (and that a disc would be lighter still because its weight is still 'more spread out'). There can be no clearer demonstration of these children's egocentric approach to weight: weight is an unquantifiable quality that affects the scales in precisely the same way as it affects the human hand. This is also what Dal suggested when he said that the fragments were lighter because 'they have been separated; they're thinner ... they're too small and then they are at the edge of the pan'; or what Gai meant when he said that the ball 'is in one piece', while 'you can see that the small bits are thin'. For others, on the contrary, the fragments were heavier than the whole 'because there are more pieces'. But here, once again, the failure to equate the whole with the sum of its parts is not due to a lack of logical resources — after all these resources are fully deployed during the quantification of matter. The reason is rather that the egocentric character of the quality 'weight' stands in the way of real operations, i.e. of reversible co-ordinations. Thus Mel argued that 'the small bits are heavier, because they're all over the pan', and that the ball was heavier because it was 'right in the middle'. This was simply Got's argument but in reverse: by relying on subjective impressions the child can arrive at incompatible theses with equal ease. Ado even changed his mind, within a matter of minutes: while he first said that the red pieces were heavier 'because they're all spread out', he went on to assert the contrary, i.e. that the red pieces were heavier 'because the blues are more spread out'!

But might it not be possible that all these interpretations, however false, are completely logical in themselves, and that they could be fitted into a formal system of coherent and reversible operations? Thus the proposition 'spread out = heavy' might serve as the basis of a 'grouping' of relations in which increases in weight correspond to increases in area, the inverse operation being based on the relation 'concentrated = light'. This is precisely what Mel seemed to say when, in answer to the question 'Can we make a ball out of them again?' he replied: 'Yes, then they'll have the same weight because we'll have a ball, and it will be right in the middle. The little bits take up more room and so they are heavier.' However, quite apart from the fact that Dal, Gai, Got, *et al.* would then have been constructing the inverse and contradictory grouping, it must also be pointed out that spatial dispersion was not the only criterion of weight these children employed. Thus had the dispersed fragments been made thinner and thinner Mel would have argued that the total weight had decreased, because earlier (see the beginning of

§1) he had claimed that the ball was heavier than the coil because 'when it's in a ball the clay is all squashed', while the coil was lighter because 'it's more spread out'! Hence when he granted that it was possible to return to the starting point he was simply thinking of an empirical return, not of real reversibility.

Here we have the true reason for the logical irreversibility of the child's perceptive weight relations, and it is worth stressing this fact because it is highly characteristic of the systematic obstacles he must overcome in all spheres before he can quantify physical qualities: the subjective relation 'spread out = heavy' cannot be constructed into an indefinite series, because once it has gone beyond a certain limit it ends up in a contradiction: the clay is the heavier the more it is scattered, until eventually it becomes lighter because it is more 'spread out'. (Children who start out with 'spread out = light' have the same difficulties, for instance Got when he was asked to compare the disc with the separate pieces.) Now these contradictions arise because the initial relation employed by the child involves heterogeneous elements, i.e. elements that are both subjective and objective, and that, being undifferentiated, cannot be co-ordinated. Since, moreover, every physical relation such as weight is complex, i.e. the product of a logical multiplication of simple relations, it cannot be 'grouped' without a prior dissociation, and this dissociation is impossible when the final relation is confused with the undifferentiated initial one. Progress in understanding therefore depends on an advance from egocentrism to 'grouping', i.e. on a dissociation of the self from the objective data constituting the *conditio sine qua non* of the grouping and also its result: only objective relations can be fitted into operational systems open to indefinite and reversible composition. The reason why children at subsequent stages dismiss as utterly absurd the idea that weight changes with expansion or division is not, therefore, that they consider it empirically impossible (rather, such changes agree with their direct experiences) but that the relations involved in weight cannot be grouped quantitatively unless their product is left unchanged.

§2. *The second stage (Stages IIA and IIB) continued: the non-conservation of weight and movement*

Before dealing with Sub-stage IIB we must first look at another aspect of the non-conservation of weight, namely at the connection between weight and movement. If primitive ideas of weight are indeed bound up with such subjective qualities as muscular effort, then we are entitled to ask if the child will also think that a ball changes its weight when it is set in motion. To settle this point, we showed our subjects two identical balls, and after asking them to weigh them in their hands, we tied one of the balls to a string, rotated it, and asked if it still had the same weight as the other. Now while children at subsequent stages thought that it obviously had, those of our subjects who still assumed that the weight of an object varies with changes in its shape or with divisions, also believed that the weight varies directly or indirectly with motion.

Here, first of all, are three examples of children at Stage I, i.e. children who think that not only the weight but even the quantity of matter changes with motion:

ROU (4; 6). 'Here are two balls; is one of them heavier than the other?' He weighs them in his hand. *'No, they are both the same.*—Now look (rotation) do they still weigh the same?—*No, the one which keeps turning is heavier.*—Why?—*Because it's bigger.*—Is it bigger?—*Yes.*—Why?—*Because it turns.'*

DUR (6; 0). 'Do they weigh the same?—*No, the one which keeps turning is heavier.*—Why?—*Because it has more clay.*—And when both are put down on the table?—*They weigh the same.*—Why?—*They have the same clay.*—And if we spin the other one?—*It would* become heavier.'

SALA (7; 6). *'The one that stays still is the heavier one.*—Why?—*Because it has a bit more clay.*—And if we spin this one (we rotate the second ball and put the first one down) do they weigh the same?—*No.*—Why not?—*The one that stays still has a little more clay.*—Why does the other one have less clay?—*Because when we spin it, it becomes lighter.'*

Here now are some examples from Stage II, i.e. from subjects who deny the conservation of weight while granting that of substance:

RAD (6; 6) says that the two stationary balls are of equal weight, but thinks that when one of them is rotated it becomes a little heavier. 'Why?—*Because it's turning.*—And the other one?—*It's lighter.*—Why?—*It's smaller.*—But you just said they were the same?—*Yes.*—And now (rotation) do they have as much clay as each other?—*Yes.*—And the same weight?—*No.*—Why not?—*The one that keeps turning is bigger.*—Why?—*Because it's heavier.*—Why?—*Because it spins.'*

LET (6; 6). *'That one is heavier.*—Why?—*Because it turns.*—Why is it heavier when it turns?—*Because the wind pulls us along.*—What does that mean?—*It's stronger.'*

KOD (7; 0). *'This one is heavier because it turns.*—And if I put it down on the table and spin the other one?—*The other one will be heavier because it spins.*—Why is it heavier when it spins?—*Because it turns a lot.'*

FIL (7; 6). *'The one that turns is heavier.*—Why?—*You can feel it on the string.*—You can feel what?—*That it's heavier when it turns.'*

CAB (7; 6). *'The one that turns is heavier.*—Why?—*It's obvious, because the wind carries the rain.*—What do you mean?—* . . .* —But do these two balls weigh the same?—*No, that one is heavier.*—Why?—*It must be because it makes air when it turns.'*

GAN (8; 0). 'Do they still have the same weight?—*No.*—Why not?—*The one that keeps turning is lighter; it's got a string* (= is being supported).'

The importance of these replies is that they reveal once again, but in a novel manner, the undifferentiated character of the primitive notion of weight. As we have shown in §1, at the lowest stage of development, weight is the quality of what presses down on the child's hand, etc. Now motion, too, clearly affects the pressure so that the child not only identifies weight with mass – which is only to be expected – but also with all sorts of forces; moreover, these various components cannot be disassociated from one another while they remain undifferentiated from actions considered as so many qualitative functions of the child's own. This is why most of our subjects believe that the rotating ball is heavier than the stationary one, either because it 'turns a lot' (Kod), and 'makes air' (Cab), or else because 'the wind pulls us along' (Let). In other words, the rotating ball increases its weight thanks to the force it has acquired: the force of propulsion and the weight being deemed identical. As for those who believe that the weight decreases with motion, their opinion, although contradicting that of the first group, is, in fact, based on the same reasoning: 'When we spin it, it becomes lighter,' said Sala. 'It's got a string,' added Gan, meaning that its weight decreases in the same way that a swimmer feels his weight diminish when he is supported by a belt. Hence, no matter whether the weight is thought to increase because the subject evaluates it by the pressure the rotating ball exerts on his hand, or whether the weight is thought to decrease by analogy with a swimmer's, in either case the confusion of weight with mass and force is due to its reduction to the tactilo-muscular qualities of the subject's actions or impressions. In either case, consequently, progress in objective weight quantification must again be based on the construction of a reversible system in which the objective relations constitute a closed grouping that leaves the weight unchanged and co-ordinates the subjective impressions in the light of objective reality.

§3. The first sub-stage of the third stage (Stage III A): intermediate reactions between non-conservation and conservation of weight

As with the conservation of substance, we must again distinguish an intermediate group separating those who deny the conservation of weight from those who affirm it *a priori* as a logical necessity. By paying special attention to their vacillations, we shall gain a better grasp of the logico-mathematical mechanisms leading to the conservation of weight. Here, first of all, some responses to various deformations of the undivided clay ball:

CRU (7; 6). Ball and coil: *'It won't be the same. Oh, yes it will, because you've used the same* (clay) *as before. It's the same weight as before.*—Can we turn this coil back into a ball like the other one?—*I think it will be a little heavier; there's a little bit more here* (points to the end hanging over the edge of the pan). *When we weigh this sausage* (i.e. the ball refashioned out of the coil) *it'll be heavier because you put a little bit extra in.*—Will it weigh more (the demonstrator points to the balance)?—*No, it won't, because you haven't added anything.*—And if I turn it back into a ball?—*It will be the same as before.*—The same weight?—*I think so.'*

LIP (7; 10). Disc and ball: *'It won't weigh the same because this one is thin* (the disc), *but it's still the same* (clay) *because now it is long and before it was a ball.'*

FLON (9; 0). Ball and coil: *'They have the same weight because it is still the same ball; oh no, the round one is thicker. And if one of them is thinner, they can't have the same weight, can they? Oh yes, they can, because the thin one is longer.*—*It's still the same ball; it has just been changed round, so it must be the same weight.'*

BEN (9; 2). *'The coil will weigh more because it gets heavier when it's longer.*—Is there more clay in it?—*No, the same.*—And the weight?—*The coil is a bit heavier than the ball.*—Why?—*Because it's a bit bigger, you can see the difference when you turn it back into a ball.*—Have you seen the difference?—*Yes, I've seen it, I've put it all together in my head* (makes a gesture of reshaping the ball) *and it's come out just a tiny bit bigger.'*

Ball and disc: *'This one* (the disc) *is lighter because it's thinner and that makes it less heavy; it isn't as strong. When it's rounder or longer and sticks out, it can't pull the blue one down.*—What do you mean?—*It hasn't the force on the pan.*—What does the force depend on?—*When it's round in a ball and big it gets heavier.*—Can we turn the disc back into a ball?—*Yes, they're almost the same because we weighed them before.*—And in the meantime?—*In the meantime it's become heavier because it's longer* (he is now thinking of the diameter!).—But it'll still make a ball of the same weight?—*Of course it will; when it's in a ball again it will have to be pressed down. When it's in a ball it has more weight than when it's flat.'* The demonstrator turns the disc back into a ball, and transforms the other ball into a disc: 'Well, which one is heavier?—*Neither, because I see now that the ball is the same size so it has the same weight.'* Ben has thus discovered the conservation of weight thanks to two simultaneous inversions.

CHAN (9; 6). Ball and coil: *'This one* (the coil) *is heavier.*—Why?—*Because it's longer.*—Is there more clay in it?—*No, the same.* (Reflects.) *Oh, it weighs the same because it weighed the same before, and now it has as much clay.*—And if I turn it back into a ball will it still be the same?—*No ... Yes.*—And if I turn it into a disc?—*The weight will be the same. No, that one* (the disc) *will be heavier.*—And on the balance?—*Heavier as well.*—And if I turn it back into a ball?—*The same weight.*—Why?—*Because they had the same weight before.*—And now?—*They haven't.'*

GRA (10; 0). 'Look at this ball. What happens if I turn it into a coil and weigh it?—*It'll be the same weight...No, it will be lighter...No, heavier.*—Which do you think is right?—*It'll be the same because it is the same clay, the same amount, but it's been pulled out.'* A moment later Gra, preoccupied with his 'but' exclaims spontaneously: *'I am trying to see how we can turn that coil back into a ball. It's lighter and the ball is heavier. When it's longer it's lighter, because it's more spread out.*—And if I roll it all together?—*That'll be the same weight.'* As for the disc: *'I think it's lighter because it's thinner; it must be lighter when it's very thin.'*

SAZ (10; 6). Ball and coil: *'They weigh the same.*—Why?—*That one is long but the other one is round.*—And so?—*Oh no, this bit sticks out* (the end of the coil protruding beyond the edge of the pan) *and so there is less. The ball is heavier.'* The coil is wound into a semi-circle so that no part of it protrudes over the pan. *'Now it's the same weight.*—And if I pull it out a bit?' The demonstrator lengthens the coil but keeps all of it inside the pan. *'That's the same, it's all inside.'* The ball is turned into a disc: *'It's the same weight, it's just flat instead of round.*—Other boys have told me that it's lighter.—*It's the same, because if we turn this disc into a ball it will weigh the same: it's thinner here* (points to the thickness of the disc) *but larger, and the ball is a little smaller here* (width) *but a bit thicker there* (height).'

Here now are some intermediate reactions to divisions of the ball:

NOS (7; 6). One of the balls is divided into seven parts: *'It'll be almost the same, but it'll weigh a bit less like that because it's in small bits.*—And if we put them all together into a ball?—*That'll be like the other ball because you have taken nothing away.*—And will it be as big?—*No, smaller because you have stuck it together into one piece; it will weigh less because it'll be smaller.'* But when the ball is divided gradually and not into seven pieces straightaway, Nos grasps the conservation: 'What if I cut this ball into two little ones?—*They'll be the same weight.*—And if I cut it into four?—*It's still the same, you've taken nothing away so it must be the same weight.'* And so on for six, eight and ten fragments.

LIP (7; 10) also starts out by asserting that the undivided ball is heavier than the seven fragments: 'And if I put them back into a single ball?—*It'll weigh the same because it will be fat.'* But when the ball is divided step by step into two, four, eight, etc. fragments and he is asked to compare them to the undivided ball, the coil and the disc, Lip replies, *'They all weigh the same.'*

DIN (8; 2) says first of all (of the seven fragments): *'These little bits will be lighter because they are small,'* but adds: *'When you turn it back into a ball, it'll be the same.'* Coil and nine fragments: *'It's the same, because if we put them back into a ball, they will weigh the same.'*

CHAN (9; 6) hesitates about the division, as he previously did about the deformations, but finishes up by saying: *'It's the same thing, it's all the clay in the ball but separate.'*

SAZ (10; 6) first believes the pieces are lighter, but then says: *'It's the same weight because all of it is inside; it's the same as the ball that used to be there.'*

SAM (10; 6) takes a good look at the ball and the seven little balls and then says: *'Here we have small bits and there we have just one, but the ball had the same weight before, so I don't really know.*—Why not?—*One might say it's lighter because it's in small bits, but this one won't shift the balance* (= change the weight). *This ball has as much clay as those small bits, the same size, the same weight.'*

GRA (10; 0). *'Oh, the bits are lighter.—*Why?*— Because they are all spread out ... but if you squeezed them all together again you would get the same ball, the same weight. But now it's lighter because it's all over the place.'* Next he decides in favour of conservation, but then he wavers again.

These intermediate reactions are of great importance because they lay bare the thought mechanisms by which our subjects try to resolve the conflict between their perceptive, egocentric view of the relations involved and the rational co-ordination of the relations.

Their subjective evaluations of the weight, first of all, are identical with those of subjects at Stage II (§1) except that their language has changed (the disc exerts less 'force' on the pan than the ball, said Ben, etc.). However, in §2 we saw enough examples of identifications of weight with force to realize that no really fresh element has been introduced. All these Stage IIIA responses are, in fact, based on the same reduction of weight to the subjective sensation of pressure as we encountered at Stage II.

But how do our subjects eventually manage to discard this egocentric view to advance to objective quantification? We shall see that they do so by the self-same constructive process as went into the conservation of matter, i.e. by gradual progress towards reversible compositions of the logical or qualitative relations involved, and by the extensive quantification of these relations through the equalization of differences. However, because that equalization meets specific obstacles in the case of weight (which apparently differs according to the distribution of matter), the problem is posed in new terms, so that the transition from egocentrism to the grouping of objective relations is not a smooth continuation of the construction of the substantial invariant, but involves a new decentration of the perceptive relations.

Let us look first of all at the gradual grouping of qualitative relations, which we have already described for the substance (Stage IIA) and which we rediscover as such for the weight during Stage IIIA. In either case, this grouping can be recognized by the production of arguments based on simple identification which are its direct result, the child finding it much easier to grasp the result of operations than their mechanism. Nevertheless, in a great many cases, the mechanism, too, is understood, as our subject's responses show quite unmistakably.

Let us begin with Chan, who having first claimed that the coil was heavier than the ball because it is longer, finished up by saying: 'It weighs the same because it weighed the same before and now it has as much clay.' Similarly, when presented with the fragments, he said: 'It's all the clay in the ball but separate.' In other words, it was by identifying the final with the initial state, and moreover by basing this identity on the permanence of matter, that Chan came to accept the conservation of weight (only for a moment, moreover). Similarly, Cru vacillated between the appearances (the coil is heavier because it is longer) and identification: 'You used the same clay as before', and 'You haven't added anything.' This raises the question of why children at Stage II who realize as clearly as our present subjects that no clay has been removed or added (and who, in fact, all grant the conservation of substance) should nevertheless fail to extend this identification directly to the weight. This advance of our present subjects would be totally incomprehensible if the conservation of weight were a primary factor and not the result of a grouping of operations. In fact, identification alone can never lead to a synthesis of identity with change. Thus let us assume that the child were convinced *a priori* that something is conserved during the transformation of the ball into the coil (and we saw that this is not true even of matter at Stage I). Now, not only does this realization not suffice to convince him that the weight is conserved while the shape changes (we saw that, at Stage IIB, he believes the contrary) but also, and above all, if the conservation (identification) is not considered the result of grouping, the identity and the transformation cannot be reduced to each other and their fusion becomes incomprehensible. This explains Cru's vacillations: 'This sausage will be heavier because you put a little bit extra in (= elongation)', but it does not weigh any more 'because you haven't added anything'. Now every transformation is, in fact, both 'something extra' and also 'the same thing', and the problem is precisely to discover how the child synthesizes the two. The entire work of Emile Meyersohn, so admirable for its philosophical courage, shows clearly that this synthesis cannot be made while mental activity is confined to identification and change to experience.

Does reversibility suffice where identification has failed? The answer depends, first of all, on the nature of the reversible action: a direct empirical return to the starting point does not guarantee conservation, a necessary condition of which is the grouping of operations (reversible composition). Now the latter depends on the psychological context, i.e. on whether reversibility is suggested by one of the demonstrator's questions or by the empirical facts, or whether the child himself grasps it spontaneously as a *conditio sine qua non* of the operations leading to the transformation of the ball. Now our Stage IIIA responses throw fresh light on all these questions and, more particularly, allow us to follow the child's advance from empirical reversals with their unco-ordinated relations to operational (i.e. complete) reversibility, and so to the co-ordination of all relations engendered by the appropriate operations.

Let us look again at Chan, who, after hesitating about the coil, denied that the disc had the same weight as the ball but nevertheless contended that it would recover its full weight once it was turned back into its original form. Now the reason why the possibility of an empirical return failed to convince him of the conservation of weight was, on the one hand, that he still found it problematical (he variously affirmed and denied it) and, on the other hand, that he completely ignored the co-ordination of the relations. Cru and Gra produced similar reaction, but Nos had made slight progress towards operational reversibility: he began by denying the conservation (during the division of the ball into seven pieces), but when the ball was divided more gradually into two, four, six, eight and ten segments, he suddenly grasped the operation and hence discovered the invariance.

Ben made more spectacular progress. He, in fact, tried to settle the problem of reversibility by a true mental experiment: 'I've put it all together in my head,' he said (gesturing the reconstruction of the ball). However — and this shows clearly that a mental experiment as such is never a logical argument — his inner experience persuaded him to deny the reversibility all the same: 'I have seen it (the difference); I've put it all together in my head and it's come out just a tiny bit bigger.' He nevertheless granted that it was possible to turn the disc back into the original ball: 'Yes, they're almost the same because we weighed them before.' Only — and this distinguishes the possibility of an empirical return from a reversible grouping — he did not grant the conservation of the weight straightaway: while the disc remains a disc, it is 'heavier because it's longer'. Now the proof that an empirical return differs from operational reversibility in essence and not merely in its results is the fact that Ben discovered the latter soon afterwards and this thanks to a new fact: while turning the disc back into a ball the demonstrator also transformed the second ball into a disc so that Ben was forced to view the two transformations simultaneously. Now, far from being confused, Ben was so impressed by the double transformation that when he was asked which was heavier, he immediately exclaimed: 'Neither, because I see now that the ball is the same size, so it has the same weight.' This sudden conversion, this sudden clarity ('... I see now!') shows unequivocally that Ben had not grasped the idea of reversibility until that very moment. Only when he was shown the two inverse operations simultaneously did he come to appreciate the operational character of the reversibility and hence to deduce the conservation of weight as a matter of course.

What then stopped Ben, before this final illumination, from deducing reversibility from his realization that an empirical return was possible? It was undoubtedly his failure to co-ordinate the relations into a general grouping: thus, at the beginning of the test, he used 'big' to refer to the greater length of the coil but also to the greater thickness of the ball, a fact that reveals the indeterminate nature of the relations he employed. The responses of Saz, by contrast, show to what extent true reversibility goes hand in hand with a co-ordination of the relations, and how the feeling of necessity characteristic of complete reversibility springs from the logical mechanism underlying operations grouped by reversible composition. Saz, like Ben, began by doubting the conservation of weight: the coil was lighter than the ball because its ends protruded beyond the edges of the pan, etc. However, the argument that eventually led him to conservation was extremely rigorous and hence of great importance. 'It's the same, because if we turn this disc into a ball it will weigh the same,' Saz explained. Nor did he leave it at that (in which case he might still have been thinking in terms of an empirical return) but he also justified his remark by a true composition of the relations: 'The disc is thinner but larger and the ball smaller but a bit thicker.' By adding this (logical) multiplication of the inverse relations (increase in breadth = decrease in height and *vice versa*) Saz thus endowed his reversibility with the operational or logical character of the true qualitative grouping.

True reversibility always goes hand in hand with the co-ordination of the relations constituting the compositions of groupings whose reversibility ensures operationality. These compositions are also clearly at work in the solutions offered by Flon and Lip. Flon's lengthy reflection is particularly instructive. Having first asserted the conservation of weight in the case of the coil ('it's still the same ball'), he started to waver, remembering that the ball was 'thicker', and that the 'thinner' coil ought to be lighter. But then he also remembered that 'the thin one is longer', whence his final conclusion: 'It's still the same ball; it has just been changed round, so it must be the same weight.' Lip similarly declared that the disc weighs as much as the ball because, though it is thin and light, it is also 'long', i.e. its length offsets its thinness. In other words, with the help of the grouping of operations and of the relations they engender he was able to fuse identity with change, the first being guaranteed by the reversibility of all the transformations, and the second appearing not only as an empirical datum but also as a result of the composition of all the possible combinations.

Now this co-ordination of qualitative relations in the same kind of grouping we encountered during our discussion of the conservation of matter is immediately extended into quantifying operations of an extensive or metrical kind. In the case of weight, these operations reflect new conditions but their formal structure is identical with those at work in the conservation of substance. Thus, in formal respects, the operation enabling the subject to assert the conservation of weight is the same operation we met at the end of Chapter 1 under the name of the equalization of differences (methods 2 and 4). However, while it is relatively easy to think of the substance of the clay ball as being made up of parcels that are simply displaced in the course of deformations without being changed in themselves, it is much more difficult to apply this schema to weight: children at Stage II still think that one and the same parcel changes weight during displacements and that the total pressure it exerts depends on its position. The specific problem posed by the quantification of weight is therefore to render the parts of the whole homogeneous, in other words to construct units despite the qualitative differences. The grasp of the conservation of weight thus depends on the realization that the weight has a homogeneous distribution and also that it is possible to divide the substance into equal parcels whose sum will weigh as much as the whole. Now this is precisely what our intermediate subjects find so difficult to comprehend in the case of divisions, as witness their progress from 'the bits are lighter because they are all spread out' (Gra), to the final discovery: 'it's all the clay in the ball, but separate' (Chan).

How was this solution arrived at? Three related factors seem to have been at work, chief amongst them the contradictions to which the subject is led by the composition of the subjective relations: thus Gra vacillated between the coil is 'lighter' and 'no, it's heavier' whence he concluded that 'it's the same'. The second factor is the subsequent discovery of the purely subjective character of his relations, as a result of which Sam, for instance, came to dissociate his subjective impression ('One might say it's lighter because it's in small bits') from the objective reading of the balance ('but it won't shift the balance') – a clear sign that he was beginning to lose faith in his egocentric evaluations. Third, and decisively, the weight now divorced from the perceptive intuition becomes attached to the object itself, i.e. its quantification is fused with the conservation of matter. This is what we shall see more clearly in the following reactions.

§4. *The second sub-stage of the third stage (Stage IIIB): conservation of weight and of substance but not of volume*

Sub-stage IIIB sees the immediate affirmation of the conservation of weight, conceived as a logical necessity.

ROB (8; 0). Coil and disc: *'They weigh the same because they're the same size; if both were balls they would be the same.'*

JAN (9; 2). *'It's the same weight. All we've changed is the shape; if they didn't weigh the same we should have had to take some clay from one of them.'*

FOG (9; 9). *'They're the same weight. They're the same balls, you've just pulled this one out.*—Didn't the weight change when I pulled it out?—*First it was round and now it's long, but it's the same clay; you didn't take any away.*—Can I turn it back into a ball that weighs the same as before?—*Of course you can, there's no extra clay.'*

BRU (9; 10). *'It's the same; it weighs the same, it's the same clay pulled out into another shape.*—Are you sure they weigh the same?—*Quite sure, because it's the same ball of clay.*—What if I turn this coil back into a ball?—*It will still weigh the same because it'll be the same ball as before.'*

BON (10; 1). *'It's the same weight; this one is long and that one is round but it's the same weight.*—Some boys have told me that it isn't.—*This one* (the coil) *is thinner and longer and that one* (the ball) *is bigger and higher, so they're the same.'*

SER (10; 0). *'It is bigger but it weighs the same. It was just squashed together before, and now it has been pulled out, but it weighs the same.'*

DUB (10; 6). Coil: *'It's the same because when it was in a ball it was the same weight as the other one. You used up all the clay of the ball, the weight hasn't changed.'* Wad of cottonwool: *'You teased out this lot and you squashed the other, but they're the same weight.*—And if I took half of the tighter wad and half of the other?—*Half of one is the same as half of the other.*—And if I took a tenth?—*A tenth of the one will still be the same as a tenth of the other.'*

ROU (11; 0). *'It's longer but it makes no difference to the weight.*—Why not?—*Because it's longer but thinner and straighter. They're the same weight, I'm sure.'*

GEI (11; 0). *'It's the same ball, only pulled out.'*

MA (12; 0). *'It's the same weight, the weight stays the same, it's the same thing inside.*—But haven't we changed the shape?—*That's got nothing to do with the weight, it's still the same amount.'*

Now for the division of the ball:

FOG (9; 9). The ball is cut into eight pieces: 'Is it the same weight?—*Of course it is, there's no extra clay, even though it's been cut up, it's still the same.*—Some boys say it's heavier.—*When the pieces are bigger there are fewer of them, and when they are smaller there are more of them. It's all the same.'*

GIV (11; 0). Seven fragments: *'It's the same clay, so it can't be any lighter.*—But some children tell me that it is.—*But we've got a whole lot of small pieces, so it must be the same together. A large ball would take their place; it's like when you cut four slices of cake. If you weigh them and put them together again and weigh them again, you'll see it always comes to the same.'*

OUX (12; 0). *'It weighs the same, because if you put those little bits together you'll get the same ball as before.'*

And finally when one of the balls is rotated on a string (*cf.* §2):

AD (8: 0). *'It's the same because it's the same clay. It makes no difference if you spin it.'*

We can now solve the problem we left undecided at the end of §3 (Stage IIIA), i.e. that of the relation between the conservation of weight and the conservation of substance, and especially between the partition and additive composition of weight and substance respectively.

Quite plainly subjects at Stage IIIB use the same arguments as those at Stage IIIA, but they complete them with an apodictic affirmation of the conservation of weight. Thus Fog, Bru and Dub said straightaway that the weight had remained unchanged because no clay had been taken away or added (identification). Rob, remarking on the overall shape, and Giv and Oux commenting on the division, spontaneously invoked reversibility, and Rou, Bon, Ser and Fog defined the composition of the relations when they said: 'It's longer but thinner' (Rou); 'This one is thinner and longer and that one (the ball) is bigger and higher' (Bon), etc. Finally, Fog, Giv and Dub produced splendid examples of quantification. Giv deployed the axiom of additive composition (that the whole is equal to the sum of the parts) when he said: 'But we've got a whole lot of small pieces, so it must be the same together.' Fog explained that the number of parts is inversely proportional to their size: 'When the pieces are bigger there are fewer of them, and when they're small there are more of them.' Finally, Dub established the homogeneity and equality of the parts regardless of their spatial arrangement: a half or a tenth of the compact wad of cottonwool is equal in weight to a half or a tenth of the loose wad.

Now while these data corroborate our analysis of the construction of operationality during Stage IIIA (§3), they also reveal a relatively new fact: the child's realization that the conservation of matter implies the conservation of weight. No doubt, this nexus was already sensed at Stage IIIA, but it must be stressed that it is only when the conservation of weight has become a logical certainty that its justifications are increasingly based on the conservation of matter. 'If they didn't weigh the same,' Jan said, for example, 'we should have had to take some clay from one of them.' This clearly showed that, to him, the conservation of matter implied the conservation of weight. Similarly Fog explained that the weight had not changed upon elongation because 'there's no extra clay'. 'It's the same clay pulled out,' Bru explained. 'You used up all the clay of the ball, the weight hasn't changed' (Dub). Above all: 'The weight stays the same, it's the same thing inside' (Ma). In short, all these children took it for granted that the conservation of matter necessarily implies the conservation of weight, while all subjects at Stage II (§1 and §2) affirmed the conservation of matter but still denied the conservation of weight. How can we explain this apparent paradox? Since the gradual construction of the links between weight and matter from Stage I to Stage III holds the key to the eventual grasp of the conservation of weight, we can do no better than conclude this chapter with a recapitulation of this process.

During Stage I neither matter nor weight is conserved, both being treated as functions of the direct perceptive relations imposed on the subject by his combined egocentrism and phenomenalism. His egocentrism reduces weight to a quality of what is being weighed or moved, and matter to a quality of what can be seen or retrieved by the eye. Now the subjective qualities of lightness and heaviness vary with shape, and though the proces of retrieval is completed at the end of the first year of life in respect of the total perceptive object, it does not yet extend to the parts of this object, i.e. to the elementary objects (qualitative parts or units) whose combination constitutes its matter or substance. Phenomenalism, in its turn, prevents these children from recomposing and grouping the perceptive relations into rational systems and thus from going beyond the appearances.

During Stage II, the logical conservation and quantification of matter are completed. Just as the infant comes to appreciate that solids *qua* perceptible objects can be retrieved as such even when they have left the visual field or when they have apparently become distorted, thus decentring his approach thanks to the construction of a group of spatial displacements (displacements of the object or of the child's body), so the growing child eventually discovers that, in respect of matter too, the parts of a deformed object can be retrieved intellectually by ignoring its links with subjective perception, and this thanks to a logical grouping of the relations defining the deformation. However, while the invariance of matter is thus constructed with the help of decentration, weight, by contrast, still remains steeped in egocentrism and phenomenalism. In fact, the substantial parts of the object are not yet considered to be of homogeneous weight, because the child still believes that their distribution affects their pressure on his hand. The conservation of matter therefore does not entail the conservation of weight as a matter of course, the dissociation between the two being most pronounced at Stage IIB.

During Stage III, by contrast, the conservation of matter comes to entail that of weight. To explain this sudden reversal, we need merely postulate that the action of weighing, too, has been decentred from the ego to become part of an operational grouping that renders it objective — with some delay, precisely because it poses greater perceptive problems. Now this new decentration, or this new victory over egocentrism and phenomenalism, is facilitated by the prior construction of the substantial invariant: apparent changes in weight are henceforth attributed to purely subjective impressions while the external relations are fitted into the appropriate operational framework. The grouping ensuring the conservation of matter is accordingly extended to the conservation of weight, each unit of matter being assigned a constant weight, so that the total weight can be obtained from the addition of these now homogeneous units, much as the total object can be obtained from the recombination of its parts. As a result, the relations between substance and weight, which were fused in the initial egocentrism and phenomenalism, later to become dissociated once again, are at long last combined into a single rational grouping. All that remains is to discover how this grouping is eventually extended to include the conservation of volume.

CHAPTER THREE

The Conservation of Volume at Equal Concentrations of Matter

In Chapter 1 we saw that the concept of substance or quantity of matter is undifferentiated. When substance is not yet conserved (Stage I) the child confuses it with volume and weight: to justify his view that the ball increases in substance, he will say indifferently that it becomes larger or heavier. By contrast, once he has grasped the conservation of matter but not yet of weight (Stage II), he begins to differentiate it from the conservation of weight, and of course, from that of volume as well. Thus while he originally applied 'large' to substance or volume indifferently, he now uses this term in two distinct senses. Let us look at this development more closely.

Our first problem is one of terminology: to test the child's conception of volume we cannot simply introduce verbal distinctions between 'large' and 'big' or add new questions to our questionnaires. Trial and error has, in fact, convinced us that the best means of assessing the child's approach to volume is to rely on the amount of water displaced by a clay ball, etc. The water level before immersion of the ball is marked in ink or with an elastic band and the child is asked whether the water level will rise, remain as it was, or fall once the ball has been dropped in. This first question contains a deliberate suggestion: we found elsewhere[1] that a number of children between the ages of five and eight years are surprised to see the level rise upon the immersion of a solid, and that most subjects under the age of eight or nine years explain the rise in level not so much in terms of the volume as of the weight of the solids immersed, which according to them produces an upward current. Now these two reactions — failure to anticipate the displacement and confusion of volume with weight — have a direct bearing on our study of the conservation of volume, as we shall see in this chapter and also when dealing with the dissolution of a lump of sugar (Chapters 4–6).

[1] J. Piaget, *The Child's Conception of Physical Causality*, Routledge & Kegan Paul, 1930, Chapter VII.

But let us return to the test. Once the subject has noted and marked the new level, he is shown another ball of the same shape and size and is asked how high the level will rise with that one. (All the children whose responses will be quoted in this chapter pointed to the correct level, thus showing that they had fully grasped the data.) This done, the second ball is transformed into a coil, disc, etc. or divided into pieces, the child being asked every time: 'And now, will it take up the same space in the water as the first ball? How high will the water rise?' etc.

With the help of this technique we were able to establish that the conservation of volume is constructed at a later stage (Stage IV) than the conservation of weight (Stage III). Thus while the child at Stage III grants the invariance of matter and weight but still believes that the volume varies with every deformation of the ball or with every division, at Sub-stage IVA, by contrast, he begins to accept the conservation of volume in certain cases, and at Sub-stage IVB in all cases.

It should be stressed that the concept of volume examined here is a physical notion, not merely a geometrical one: the volume of the ball is the space occupied by a substance thought to be both impenetrable and incompressible or, at the very least, to remain at equal concentration throughout the experiment. That this concept is complex goes without saying, and it is for this very reason that the conservation of volume appears at a later stage than the rest: it is not until the age of eleven or twelve years that physical volume is treated as an invariant on the same footing as substance and weight, and that its composition with weight gives rise to the notions of density or of compression and decompression.

§1. *The third stage (Stages IIIA and IIIB): conservation of substance and weight and non-conservation of volume at equal concentrations of matter*

We shall first look at subjects who do not yet grasp the conservation of volume both during deformations and also during divisions of the clay ball, beginning with one or two examples at Stages I and II (for comparison):

ADA (6; 0) does not believe in any type of conservation (Stage I). As far as the volume is concerned, he thinks that when the ball is divided into six *'the water will rise less because they are such tiny little bits.*—Is there as much clay in them?—*There is more clay in the ball, it's heavier.*—And if I turn them back into a ball and change the other ball into a coil?—*The ball will make the water rise because it's more rolled up* (= in a piece).'

ROD (7; 0) accepts the conservation of matter but not that of weight (Stage

IIB). Ball and coil: *'It goes higher with the round one because it's bigger so it makes it go higher.'* The ball is divided into five and the coil refashioned into a ball. *'It'll go lower with the pieces. They are thin and take up less space.—*Do they have as much clay as the ball?—*Yes.—*Are they as heavy?—*The small pieces weigh less.—*And how about the space they take up?—*They'll take up less room.'* The experiment is performed and Rod notes that the water reaches the same level. *'Ah yes, all the same, the small bits together are the same as the whole ball.'*

NAL (7; 6), also at Stage IIB, predicts that *'the water is going to rise, the ball will take up some of its space.—*And if I put the bits in?—*The water'll rise a little less, because they are small bits, and that's not the same as the ball.—*Why not?—*The ball is in one piece.—*And if I make a sausage out of the bits?—*It'll rise higher than with the ball. Before it was shorter, but now it is longer.'* The coil is changed into a ball and the second ball into a coil. *'This one* (the new coil) *will make it go higher because it's bigger.'*

Next we shall quote the responses of two subjects at Stage IIIA, i.e. children who believe unreservedly in the conservation of matter but produce intermediate reactions in respect of the weight:

BEG (9; 2). 'What will happen if I put this ball into the water?—*The water will rise, because of the clay in it. When I put my hands in the water I make it rise as well, because my hands take up room in the water.—*And what about this coil (second ball transformed)?—*It'll rise a little more, because that sausage is long so it takes up more room.—*Does it make any difference if I put it in this way or that (horizontally or vertically)?—*Like this* (horizontally) *it'll take up a little more room, like that* (vertically) *it'll take up a little less.—*Why?—*...—*And when it is in pieces (the coil is divided into five parts)?—*It'll take up less room because it's smaller and lighter.—*And if I put them back into one piece?—*Then it will be the same as before.'* Disc: *'That'll take up less room because it's flat and can lie at the bottom of the water; it's thin.'*

CLAD (10; 5). Stage IIIA: *'The water will rise because of the weight; if you put stones into water they take up room.'* Coil: *'It'll rise less; that roll is looser.—*But why will it rise less?—*The roll takes up less room; it goes more into the corners while that one* (the ball) *stays right in the middle.—*And on the scales?—The roll will take up more room.—*And in the water?—*The ball.—*Why?—*It's large and round, and the other one is pulled out and quite thin.—*And in pieces?—*When it's in small pieces it'll take up more room; the bits go into all the corners but the water will rise less because they're lighter.'*

Here finally are some responses by subjects at Stage IIIB, i.e. by subjects who are convinced of the conservation of matter and weight in all circumstances but who, like the preceding subjects, deny the conservation of volume even at equal densities:

MEY (8; 0, advanced) thinks that if a cylinder of clay is immersed horizontally it will cause the water to rise to a certain level. He is shown an identical cylinder: 'Has it the same amount of clay?—*Yes, it's as heavy.—*And if I put it into the water but lengthwise like that (vertically), will the water rise as high as it did before?—*No, a little less because it's thin and when it's *thin it takes up less room.'* The first cylinder is cut into pieces: 'Are they the same weight?— *Yes, it's the same because it's all stayed together* (=because the sum of the parts is equal to the whole).—Is there as much clay as before?— *It's the same amount but not so big.*—And in the water?— *It'll rise less because the small pieces go any which way and that takes up less room.'*

LAD (10; 6). 'What happens if I put this ball into the water?— *It'll rise. The water always rises when you put anything in the bottom. It'll rise more if it's bigger, because it's got a better chance then. It goes more to the bottom and the water rises more easily.*—And with this roll?— *The sausage will take up more space.'* The coil is cut up. *'You don't have to go on cutting it, I can guess what'll happen: the bits will make the water rise less than the ball because they are smaller.'*

GOT (11; 0). First ball: *'The water will rise because that'll take up room.*— Why?— *It's big, it'll make the water bigger and make it rise.*—And this other ball?— *It's as big as the first.*—And if I change it into a coil?— *It won't be as big, it'll be thinner and take up less room.'* Two identical cylinders, one vertical, the other horizontal: *'It'll take up more room when it's up than when it's flat. Because when it's big like that it pushes the water up and when it's straight it takes up less room.'* Two coils, one held vertically and the other the shape of a corkscrew: *'They have the same weight.*—And in the water?— *That one* (the corkscrew) *will take up more room.'*

FRE (11; 5). Ball and coil: 'Is there as much clay in the one as in the other?— *It's absolutely the same.*—Why are you so sure?— *It's the same weight because you've taken none of the clay away and you can always turn it back into a ball.*— And if I put them in the water?— *The ball will make it come up more; it's bigger and takes up more room than its weight* (= it is bigger but of equal weight). *That one* (the ball) *is large and this one* (the coil) *thin, so it'll make the water rise less.*—Look at what I am doing now (the ball is cut into seven pieces).—*Oh, I can tell you straight off it'll be the same weight.*—Why?— *There are as many small bits as in the ball and if you put them together again you'll get the same as before.*—Will they take up the same room in the water?— *The ball is bigger so more water will rise with it. Am I right?*—Think about it. Why should the ball make more of the water rise?— *These here are small bits. Oh no, there are lots of them so they will take up more room.'* The pieces are joined together into a disc: *'I can guess what'll happen. This one is flat and that one is large; they're the same weight because they have as much clay as each other.*—How far will the water rise with the disc?— *Less than with the ball, this one is thin and that one is fat.*—But they have as much clay as each other?— *Yes, because if you turn it back into a ball it'll be the same.*—And in the water?— *Ah, I've gone wrong. The disc will take up more room and the ball less.*—And by how much will the water rise?— *It'll rise more with the disc. It's larger so it takes up more space.'*

These responses highlight the obstacles in the path of the conservation of the volume of solids, and explain why it is not grasped until a relatively late stage. True, our first two groups of children (Stages I–II and IIIA) tell us nothing new in this respect; since they have failed to grasp the conservation of matter and weight, it is only to be expected that they should fail to grasp the conservation of volume: if the ball seems to lose substance and weight when it becomes thinner or flatter it is natural that it should also seem to decrease in size or volume. Here we have just another example of the initial predominance of perception over intellectual operation. However it is highly interesting to compare these elementary reactions to those of subjects at Stage IIIB, because they demonstrate that the physical volume of a body is a much more complex matter than that undifferentiated quality, its substance, or that differentiated quality, its weight. In fact, the perceptive aspect of the volume seems to depend on the shape of the object no less than on its dimensions and content, while physical volume, once it has been quantified and divorced from its qualitative aspects, is treated as a relation connecting the quantity of matter with its compression or concentration. This explains why Ada thought that the pieces were less voluminous than the ball because they were small and that the ball was more voluminous than the coil because it was rolled up (= in a piece); why Rod thought that the ball was heavier than the coil because it was bigger, while Beg said that the coil must displace more water because it was longer; or why Clad asserted that the more voluminous body might indifferently be the more compact ('the roll is looser; it goes more into the corners', while the ball 'stays right in the middle') or the less compressed ('When it's in small pieces, it'll take up more room; the bits go into all the corners,' but they make the water rise less because they are 'lighter'). In short, while he continues to treat both matter and weight as variable qualities, the child vacillates in his choice of criteria for determining volumes between one dimension and the next and between the compactness or tenuity of the substance.

Now what is so striking and, in fact, poses the real problem we have to solve in this chapter, is that subjects at Stage IIIB, who accept the conservation of matter and weight, reason precisely like those at earlier stages when it comes to volume. To Mey, the cylinder was less voluminous when it was immersed vertically because it was 'thin' and so it was for Got: 'It'll take up more room when it's up than when it's flat.' Lad, for his part, thought the fragments would make the water rise less because the ball 'is bigger' and hence had 'a better chance', etc. Yet all these subjects realized full well that the weight of the combined fragments equalled that of the whole ball. Why then, if they can co-ordinate the relations defining the shape of the object so well that they can group them into invariants when it comes to weight and matter, do they not deduce the conservation of physical volume as well?

The reason is that physical volume involves an extra co-ordination,

51

namely that of the quantity of matter with the concentration of the elements. Thus, suppose that the clay expands or contracts upon each deformation of the ball. In that case a constant quantity and weight of clay would not imply a constant volume. To grant the conservation of volume the child must therefore assume that every part of the given solid occupies the same space as every other part and that it neither expands nor contracts when it changes position. Now it is precisely this equalization of the parts, grasped at about the age of seven years in the case of the quantity of matter and at about the age of ten years in the case of weight, which poses a new problem in the case of physical space. Thus, when the ball has been divided, the child has no reason to assume that the sum of the separated parts must be equal to the whole, since he believes that the latter differs according to whether it is in one piece or not, i.e. according to the arrangement of its parts. Mey left us in no doubt on this point when he explained that the total weight of the fragments was equal to that of the undivided ball 'because it's all stayed together', but that the total volume had changed 'because the small pieces go any which way and that takes up less room'; and Fre was no less specific. He, too, realized that the fragments were so many parts of the original ball: 'There are as many small bits as in the ball and if you put them together again you'll get the same as before.' But he obviously felt unable to apply the argument on which he thus based the conservation of weight to the volume, because 'the ball is bigger' or alternatively because there are lots of pieces. In short, all these children believe that the volume depends on the structure of the whole, and that it is transformed upon the separation of the parts. That is why they not only believe that the volume changes with changes in shape but also, and rather paradoxically since they have granted the conservation of matter, with changes in position: the vertical cylinder is thinner and 'takes up less room' than the horizontal cylinder (Got).

This suggests – and the rest of this book will bear out this view – that the child can only grasp the conservation of volume if he assumes that matter has an atomic or granular structure whose density is unaffected by changes in shape or by divisions. Hence the conservation of volume not only implies the conservation of matter, but also a schema of constant concentration and this is why it is constructed later than the conservation of weight and why, as we shall see, it appears at the same time as the concept of density or of the compression and decompression of atomic grains. In brief, the conservation of volume, like all the other forms of conservation, depends not only on the homogeneity of the parts but also on the fact that the parts neither contract nor expand during transformations. Now, while egocentrism and phenomenalism still persist, all changes in shape or position and all division seem to go hand in hand with changes in concentration; only an implicit or explicit atomic approach can therefore lead to the idea of the conservation of physical volume.

§2. *The first sub-stage of the fourth stage (Stage IVA) : intermediate reactions between the non-conservation and the conservation of volume*

We must again pay special attention to the intermediate cases because they highlight the vacillations, and hence the difficulties in the path, of our subjects:

PEL (9; 0). 'What will happen if I put this ball into the water?—*It will sink to the bottom.*—And the water?—*The water will rise because the water at the bottom will come up.*—And if I change the ball into a sausage?—*The water will rise higher because the sausage is longer and takes up more room.*—And like that (vertical coil)?—*That'll take up a little less room.*—And if I turn it into a cake?—*The same as the ball.*—Why?—*It takes up the same room; it's the same clay as in the ball; it just has another shape.*—And the sausage?—*Ah, it's the same as well.*—And in pieces (four)?—*It's the same thing, it's all the clay from the ball but separate.'*

DEN (9; 0). The ball: *'The water will rise because we've put something in.*—And this sausage?—*That'll make it rise the same, it's the same clay, only pulled out.*—Will it take up as much room?—*No, it's thinner; oh no, they are the same; the one is long and the other one round but they take up the same room.*—And if I cut it into four?—*That will also go to the bottom, but the small bits will take up less room than the ball, than when it's one piece.'*

LER (10; 0). The ball: *'The water will rise for sure, the clay takes up room inside.'* The experiment is performed and the level is marked. The ball is retrieved and divided into seven or eight small pieces. 'And like that?—*It'll take up more room. Ah, you can see* (the pieces have not yet been dropped into the water) *it's got bigger* (= it seems bigger). *The pieces take up more room.*—Why?—*There are more bits, and when you try to put all of them back into a ball they won't go in.*—Can't we make the same ball out of them again?—*If you squash them tight you can, but you'll have to squash them very hard.'* Coil: *'The water will also rise a little bit.*—As much as with the ball or not?—*Almost the same. There is hardly any difference, perhaps none at all.'*

DREC (10; 0), by contrast, cannot tell whether or not the coil is more volumi-nous than the ball: *'It's the same thing. It's the same ball though it's no longer so round. Oh no, the sausage will take up more room.*—And if I cut it into small bits and I put all of them in the water?—*That'll be the same as the sausage.*—And also like the ball?—*No, it's more than the ball; it's bigger. If we tried to make them into the ball it would make it bigger.*—You think so?—*If we put all these bits together into a ball the water would rise more because the ball is bigger.*—And the bits of the sausage?—*The same.*—And if we turned the bits back into a sausage?—*The same size.'*

DIV (10; 6). *'The water will rise a little bit more because the sausage is longer than the ball, so it takes up more room.*—And if we lay it flat?—*That comes to the same.*—And if we cut it up into small bits?—*That'll take up the same room; it's as if we had put the whole sausage in.*—Does it take up more or less room than the ball?—*The same. I've thought it over now. The sausage is made from the ball, it's as if you put the whole ball in.'*

VIA (11; 0). Ball and coil: *'The ball will make the water rise more because it's bigger; it's got a bigger volume.*—And if we turn it into a disc?—*The same, because it's got the same weight.*—But aren't the ball and sausage the same weight?—*Yes, but they don't take up the same room.*—And if we turn the sausage back into a ball?—*That will make the same size.*—And in small bits?—*Together they'll be the same size as the ball; it'll make the water rise as much.*—And the coil?—*Ah yes, that'll also be the same.*—Why didn't you think so before?—*Because the ball is higher and this one is longer.'*

SED (11; 0). Ball and coil: *'The water will rise the same because it takes up the same room; it's longer but thinner.*—And if we turn it into a disc?—*It's the same; no, the disc is not the same; yes, it is, it's larger but it weighs the same.*—And in small bits?—*The same.'*

What strikes one immediately about these responses is that, far from introducing a novel element, they bear a close formal resemblance to the corresponding reactions at Stages IIA (conservation of substance) and IIIA (conservation of weight).

To begin with, the hesitation of these subjects and their arguments in support of the non-conservation of volume are precisely of the same type as we encountered earlier: the coil is more voluminous because it is longer and bigger, or it is less voluminous because it is thinner; the ball is more voluminous because it is round and the disc less so because it is flat; the fragments are more voluminous because they are more numerous and more scattered, or less voluminous because they are small, etc. Moreover, the arguments invoked in favour of the conservation of volume are also precisely the same as those which served our subjects some months or even years earlier to justify the conservation of substance and weight: identification, reversible composition and quantification by equalization of the parts or the differences.

Thus Pel, Den and Div appealed to the identity of the clay for their justification of the conservation of volume: the disc has 'the same clay as in the ball' (Pel); 'it's the same clay, only pulled out' (Den); 'the sausage is made from the ball, it's as if you put the whole ball in' (Div). But if identification is such a simple matter, we are entitled to ask why these subjects did not arrive at the conservation of volume just as soon as they had discovered the conservation of substance. Reversible composition, too, recurs in all its forms. To Via, the sum of the parts is equal to the initial whole: 'together they'll be the same size', and the elongation of the coil compensates for the flattening of the ball. Similarly Sed said: 'it's longer but thinner', etc. Moreover, these compositions by the addition of parts or the equalization of relations lead to a quantification of the total volume similar in all respects to a quantification of the total substance or the total weight.

The only explanation of the time lag between the respective discoveries of the conservation of matter, weight and volume—a time lag that is the more paradoxical because the arguments leading to each discovery are precisely the same from a formal point of view—is that an obstacle specific to the concept of physical volume stands in the way of the child's logical composition of the relations involved and of his equalization of the elements or differences. Now this difficulty, which as we suggested in §1, might be due to the possible expansion or compression of the ball during deformations or divisions, is reflected most sharply in the intermediate responses we have just quoted. In the case of division, for example, some subjects agreed with Den that the 'small bits' take up less room than the ball 'when it's one piece', while others (e.g. Ler and Drec) took the opposite view. Now the second group mentioned changes in compression explicitly, and not merely implicitly like some of our younger subjects who used less logical or less skilful procedures to justify their views. Thus Ler thought that when the ball is cut up it gets bigger because 'when you try to put all of them back into a ball they won't go in' unless you 'squash them very hard'. Similarly Drec, though granting that the fragments of the coil were equal to the undivided coil, thought that the latter was bigger than the ball from which it had originated. The reason why some of these children consider the coil more voluminous than the ball is therefore perfectly plain: when the ball is pulled out, the clay is thought to expand in volume, while the quantity of matter and weight remain unchanged. In short, whenever these intermediate subjects maintain that the volume changes they assume that the parts become compressed during displacements or divisions, and whenever they come down in favour of conservation it is because they have been able to render the parts homogeneous and hence of equal concentration. To put it more simply, to younger children at the lower stages of development all substances seem unstable or elastic, expanding or contracting with every change in shape, whereas the conservation of volume demands that the parts remain at constant concentration (except under special circumstances). This, as we suggested earlier, is why the grasp of the conservation of volume appears so late: it involves an atomistic schema together with the elaboration of the relations governing the concentration or density of matter.

§3. *The second sub-stage of the fourth stage (Stage IVB):* *conservation of volume*

Let us now examine the responses of children who have come to appreciate that the volume of the clay ball is necessarily conserved during all deformations or divisions.

JAS (9; 6). Ball transformed into a coil: *'The sausage takes up the same room, only the mastic (= the clay) runs lengthwise, so the water will rise to the same level; they have the same volume.*—And if we cut it up?—*It's still the same; it's the same amount of clay.'*

BUR (9; 10). Coil: *'It's just a long ball, so it must make the water rise the same; the long ball is the same as the round one.*—And if we cut it up?—*It's still the same as the ball, so the water must rise the same as before.'*

HER (10; 0). Fragments: *'You are putting in almost as much as with the ball so the water must rise; it'll rise to the same height because it's the same amount of clay.'*

VIQ (10; 6). Coil: *'It'll make the water rise as much.*—Why?—*Because it has the same weight; what I mean to say is it will take up the same room.*—Why?—*If I turn it into a ball it will be shorter but higher; when it is long, it is bigger (= higher) but flatter (= shorter), so it takes up as much room.*—And in bits?—*It will rise as much as with the ball; the bits are small but they are the same size. After all the two balls we started with were the same.*—And if we turn the coil into a disc?—*It's still the same, it's round but it's flat.*—And like this (coil held vertically)?—*It's the same, only standing up.'*

DUB (10; 10). *'The ball will have to take up some room, so the water will rise.*—And the disc?—*You've changed the shape, but it will make the water rise just as much, because it's the same weight and takes up the same room.*—And the coil?—*The water will still rise as much because there is as much clay as before.*—How can you tell it will take up the same room?—*It's obvious—it's the same amount of clay.*—But will it have the same volume?—*Yes, because it is the same amount of clay, so it must take up the same space.*—And if we turn it into a ring?—*It's still the same volume because it's the same weight.*—And in pieces?—*It's the same thing, it will take up the same space, because it's the same amount of clay and the same weight.*—How do you know?—*Because I've seen that all the clay has been kept.'*

BIV (11; 0). *'Why does the water rise?*—*Because the ball takes up space in the water.'* Eight fragments: *'The water will rise as much as before, because it's the same amount when it's in small pieces; that one* (the ball) *seems to be a little bigger but it ought to take up the same space.*—What do you mean by "it ought to"?—*Because if we made a ball with that lot we would get the same ball.'* Coil: *'It comes to the same.'* Four small coils: *'You can cut it up as much as you like, but when you put them together again it comes to the same as the ball.'*

ROUG (11; 6). Coil: *'It'll take up the same room, it's thinner but nothing has been added. It's longer but it's thinner.'*

HER (12; 0). Disc: *'It will take up the same room in the water.*—If I squash this packet of cottonwool will it take up the same space?—*No, less.*—And what about its weight?—*It will stay the same.*—And if I stretch this ball of clay will it take up the same space?—*Oh, yes, it's longer but it's thinner.'*

Just as the replies of subjects at Stage IVA were formally identical to the intermediate reactions to weight and substance (stages IIIA and IIA), so also are these justifications of the conservation of physical volume (Stage IVB) similar in all respects to the justifications of the conservation of weight and substance (Stages IIIB and IIB). This brings us back to the discussion of the mechanism of conservation in general begun in Chapter 1.

What strikes one most about these Stage IVB responses is that, despite the complex developments that prepared the way for them, they are so extremely simple. It seems that the child encounters no special problems here: the conservation seems so obvious to him that one might easily think he never doubted it, and this demonstrates how much the *a priori* is the result of a long genetic process of maturation. In other words, it represents an end point and never a beginning. Logically, all the arguments used by these subjects seem to reduce to pure identity, and this brings us back to the problem of identification—a problem whose solution, as we now realize, concerns all the four major stages we have distinguished so far.

These responses, in fact, reflect a double identification: an intrinsic identification of the different parts of the whole (the fragmentation or deformation of the whole neither diminishes, nor adds anything to, the volume, weight or substance), and an extrinsic identification of the conservation of the substance with that of the weight and the volume, so much so that the child uses any one at will to justify any of the others. Now that we are familiar with all aspects of this development, we have no difficulty in seeing that both types of identification result from a grouping of operations, and that without this reversible operational composition they would have no sense at all.

The intrinsic identification of the parts may be of a logical type (qualitative equality) or of a mathematical type (quantitative equality). To justify the conservation, the child can base his arguments either on the elements of which the whole is composed (the fragments) or on their relations. In the first case, he can simply try to establish that the displacement of a given element leaves that element unchanged (method 1: logical identification of the elements or of classes of elements); or else he can consider all the elements equal to one another and hence constituting so many units (method 2: numerical identification or mathematical equality of units). Again, if he proceeds by the co-ordination of the relations, he can (in simple cases of division, for example) confine himself to the discovery that the overall relations remain identical (the sum of the lengths of the sections is equal to the length of the undivided coil, etc.) and so apply the method of logical identity or of the qualitative equality of the relations (method 3). Finally, he may realize that the proportions remain the same, and accordingly reduce the relations to common quantitative standards (method 4: the equalization of differences or the mathematical equality of the relations). Now these are, in fact, the four methods described at the end of Chapter 1.

Thus when Jas declared that 'the sausage takes up the same room, only the mastic runs lengthwise', or when Dub said that the fragments occupy the same space as the coil 'because I've seen that all the clay has been kept', it is clear that they were applying method 1: though shifted, the fragments had remained unchanged. When Her said of the fragments 'you are putting in almost as much as with the ball', or when Viq said of the vertical disc 'it's the same, only standing up', they were applying method 3: the height of the disc had become its length and the length had become its height, the combined dimensions of the sections being identical with those of the original ball. When Biv said: 'You can cut it up as much as you like, but when you put them all together again it comes to the same thing as the ball'; or 'It's the same amount when it's in small pieces', he was completing method 1 with method 2, i.e. employing a composition that, as we shall see in the next few chapters, will eventually be extended into explicit atomism. Finally when Roug said of the coil 'it's longer but it's thinner', and when Viq and Her said much the same thing, their equalization of the differences led them from method 3 to method 4. In short, all these subjects resorted to the logical identification (qualitative equality) of the elements or relations and to their fusion into a mathematical identification (quantitative equality) of units of number or size.

However, it is clear that these identities can be established only in conjunction with operational groupings of the whole, whose existence is ensured by the inverse operation: 'If we made a ball with that lot we would get the same ball' (Biv). This poses the question of whether the operation derives from the identification, or whether the converse is true. In fact, the distinction is a purely verbal one, for we saw that the real problem involved in the construction of each new invariant is whether or not a fragment of the ball remains unchanged when it is displaced. If it preserves its material identity then the total substance is conserved; if it exerts the same pressure on the balance then the total weight remains constant despite the deformations; finally if it neither expands nor contracts, the total volume remains constant. But it is precisely the identity of the fragments upon displacements which is in question in all three cases, and that identity, as we have seen all along, is increasingly difficult to establish as the child proceeds from substance to weight and finally to volume.

What then is the precise relationship, in all three cases, between identification and operation?

Let us first of all distinguish logical or arithmetical from physical operations, the latter involving partitions and displacements in space and time, and the former the substitution of concepts or numbers for space, and of deductive for temporal sequences. Now logical operations involve both identity and change: they consist of transformations, but transformations relative to constants. Thus if $A + A' = B$ (whence $A = B - A'$ and $A' = B - A$), then A, A' and B are constants whose transformation is expressed by their addition ($+$) or subtraction ($-$). Now how do we know that A, A' and B are constants, i.e. that $A = A$, $A' = A'$, and $B = B$? Because they constitute so many 'identical operations' in the additive grouping in which they appear:[1] $A + A = A$, $A' + A' = A'$, and $B + B = B$. Much as the additive operation is impossible if the terms are not identical, so the identity cannot be established without the operational system of which it forms a part. Can we say that the child grasps the identity $A = A$ intuitively before the emergence of operations, purely by contrasting it with the difference $A + A'$? But this constant merely means that, in the equality $A + A' = B$ we cannot substitute A for A' because in that case we would arrive at the absurd result that $A = B - A$, whereas A can always be substituted for itself as can $A + A'$ for B. The identity thus results from the so-called 'identical operation', which has no precise meaning except as a function of an overall grouping.

These remarks apply *a fortiori* to the case of numerical or quantitative identity, in which case the iteration $1 + 1 = 2$ replaces the tautology $A + A = A$. For what else is the identity of the unit $1 = 1$, if not the fact that in additive or multiplicative groups of numbers any unit whatsoever can be substituted for any other?

Let us now pass on from logical or arithmetical to physical operations. Consider an empirical transformation such as the deformation or division of our clay ball. Every transformation is reflected in qualitative changes which the young child assesses in egocentric *cum* phenomenalist terms. At first, the ball strikes him as changing in volume, weight and even in quantity of matter, but when he tries to group these relations, as he is forced to do by the very contradictions to which he is inevitably led in the absence of a system, he must satisfy a condition common to all groupings, namely define the transformations as functions of invariants and

[1] See J. Piaget, 'Le groupement additif des classes', in *Proceedings of the Société de Physique et d'Histoire naturelle de Genève*, 1941.

vice versa. Now, a semblance of invariance appears on the intuitive plane thanks to the perceptive discovery that it is possible to return to the starting point. Thus an elastic band changes its volume when it is stretched, but resumes its initial bulk immediately afterwards. Similarly, a ball of clay seems to expand when it is converted into a coil, but can be restored to its original shape by the opposite action. However, though this discovery leads the child to operations, he must still surmount one crucial obstacle: so long as empirical transformations are simply discovered and not yet constructed, they look like so many creations *ex nihilo* and the opposite actions as so many acts of destruction. Before the final state can be recognized as the necessary result of the original one and *vice versa*, the child must first realize that the two are both identical and different, and that the difference can be expressed by a (+) or a (−). This introduces an operation whose invariants are constructed with the terms they engender, the transformation being defined by the operational act as such. Now because the perceptive changes do not lend themselves to being treated as homogeneous increases or decreases, and also because the objects of perception as such cannot be reduced to one another or composed by equalization, the reversible operation cannot be constructed until after the replacement of the sensory object and quality with a rational object and a rational relation. When this has happened, the invariant terms of the operation will have become equalizable elements of the object, and the operational act will be a purely spatio-temporal transformation. The perceptive qualities will have disappeared from this operational grouping and will henceforth be attached to the subject. A physical operation, therefore, is a reversible transformation like a logical or arithmetical operation, but one in which the additions, subtractions, multiplications and divisions of classes, numbers and relations are replaced by partitions and displacements in space and time, and the classes or numbers by grains or particles that can be composed with one another thanks to these spatio-temporal relations.

Let us finally look at the second type of identification which appears to be at work in our Stage IV responses, i.e. at the child's identification of the conservation of matter with that of weight or volume. In fact the conservation of volume is related to the other two conservations in the same way as that of weight is related to that of substance (see Chapter 2): mutual implication in non-conservation followed by the separate construction of the three invariants and finally by implication in conservation. Thus, at Stage I, the child generally attributes the apparent changes in volume of the ball to changes in substance and weight, or *vice versa*. At Stage II, by contrast, he no longer justifies changes in volume by changes in substance, but by changes in weight. At Stage III the invariance of weight and substance are again thought to imply each other but not the invariance of volume. Finally, at Stage IV, all three invariants have been correlated. Thus, Jas justified the conservation of the volume by that of the substance when he said that the combined fragments had the same volume as the entire coil because 'it's the same amount of clay', while Dub justified it by the conservation of weight when he said of the ring: 'it's still the same volume because it's the same weight'. Dub even went so far as to say: 'It'll take up the same space, because it's the same amount of clay and the same weight', thus clearly relying on extrinsic identification. Let us now try to elucidate the mechanism at work.

Two problems arise in this connection: why is there a time lag between the construction of the three invariants if the grouping of all the physical operations we have just defined appears at Stage II, and why is there a final fusion of the three? Before we can answer these questions, we must first define the relation between subject and object in the construction of groups of operations, which takes us back to the general problem of the transition from primitive egocentrism to grouping which we have raised in the last chapter. To begin with, we must be careful not to mistake the construction of groups of operations for the mere addition of reason to perception: not only must reason keep correcting the perceptive data and complement the apparent world with a more profound one, but it must also keep correcting the egocentrism of the subject's own perspective in such a way that the reconstruction becomes a form of decentration or conversion – a sort of Copernican revolution on a minor scale, which robs the initial reference system of its privilege, and incorporates it into a set of objective groupings.

For a better grasp of the general nature of this process, on which all the correlations we shall be examining in Chapters 4–12 depend, let us briefly recall the construction of the first invariant elaborated by the child's practical intelligence: that of the object of perception. At the beginning of mental development, sensori-motor objects seem to keep changing shape and dimensions, and sometimes even to destroy themselves when they transcend the limits of the perceptive field. During the second year of life, by contrast, the child begins to grasp their conservation. Now we have shown elsewhere[1] that this construction is identical with that of space as a whole: by fitting the successive impressions of the moving object into the

---

[1] J. Piaget, *The Child's Construction of Reality*, Routledge & Kegan Paul, 1955, Chapters I and II.

'displacement group' described by H. Poincaré, the child manages to co-ordinate them into a constant object. But before he can do so, he must first decentre the space of his own activity and fit the latter into a 'grouped' set of movements: only then does the object become detached from the subject, who has become just one element in the universe he constructs. The transition from radical egocentrism to the objective group may therefore be said to resemble transition from geocentrism to Copernican heliocentrism.

Now the construction of the three invariants (matter, weight and volume) makes us privy to the continuation of this general process, or rather to its repetition on each new plane of activity. Phenomenalist egocentrism returns at Stage I where the child refuses to believe that matter is conserved during the deformation of a clay ball. However, it is no longer the whole object whose constancy he now doubts, but only that of its fragments. At Stage II, by contrast, he decentres his approach and dissociates the subjective or apparent elements from the external reality, i.e. from the grouping of physical transformations that leave the quantity of matter unchanged. Hence, in much the same way as he has come to appreciate that his own position and movements are responsible for the apparent changes in the shape and dimensions of what is, in fact, a constant object, so he now realizes that his own perspective of the ball in no way alters the actual quantity of matter it contains, and that his perception of each change, for example the elongation of the ball into a coil, must be corrected by his perception of complementary changes. However, at Stage II, dissociation between subject and objective reality is still confined to the substance and does not yet imply a similar dissociation in the more complex sphere of the perception of weight: the child continues to believe that the weight alters with changes in perception. But because of the contradiction to which he is led by the co-ordination of the relations he has established, he is impelled at Stage III to group the weight relations into an external system which thus links up with that of the substantial relations. He accordingly realizes that each constant fragment of the substance has a constant weight and that the apparent changes are due to subjective factors, and so comes to appreciate once again that the perceptive changes result from a special perspective that must be corrected by co-ordination with other perspectives. For example, he discovers that though the disc seems lighter than the ball from which it has been made, because its weight is more 'scattered', the balance belies this conclusion, so that the subjective impression (that the more dispersed a substance is over the palm the lighter it is) must be corrected. But once again this dissociation of the subjective from the objective elements has no immediate effect on his perception of the volume because the latter depends on other factors. This explains why the phenomenalist and egocentric approach to volume persists at Stage III and why the dissociation we have described only appears at Stage IV, when changes in volume become grouped like the rest, and when every fragment of the substance is seen to conserve not only its weight but also its volume, and all apparent contractions or expansions of the whole are attributed to the subject's own perspective.

We now see why, at Stage IV, the conservation of volume is justified by the conservation of matter or of weight and *vice versa*: in all three cases, the child must express the properties of the total object by those of its elementary parts which can be grouped by the physical operations of displacement and return to the original position. This apparently simple identification is therefore also one of a great complexity, because it calls for the co-ordination of three operational groupings constructed separately. This will become even more apparent in the following chapters which deal with the child's progress from conservation to atomism.

PART II

From Conservation to Atomism

The Destruction of Matter and the Dissolution of Sugar

Our study of the concepts the child develops for dealing with the deformation of the clay ball has enabled us to follow the genesis of the conservation of substance, weight and physical volume step by step. In the next three chapters we shall be re-examining the entire problem with the help of a new experiment: the dissolution of a lump of sugar. Needless to say, we shall encounter a number of familiar reactions, but the questions they raise will be new. Thus while the clay ball merely changed its shape, dissolution constitutes a change in the state of matter and hence a much more profound transformation. Moreover, when sugar dissolves it seems to do a sort of vanishing act, and when we ask the child whether it is nevertheless conserved we are demanding a much greater mental effort from him and, in any case, an entirely different intellectual construction. In particular, the three concepts of conservation seem much more closely interrelated than they were with the clay ball. Thus during deformations of the clay ball it is simple to draw the child's attention separately to the substance ('Has something been taken away?'); to the weight ('If nothing has been taken away, does the transformation change the weight?'); and to the volume ('All other things being equal, does the elongation of the ball involve its overall expansion or contraction?'). By contrast, when one or two lumps of sugar are dissolved in a glass of water, the only experimental datum to which the child can refer is the fact that the water level does not change during and after dissolution. That being the case, does he still construct the three principles of conservation and, if so, does he construct them simultaneously or successively and in the same order as he did with the ball of clay, and why?

Finally, and above all, the dissolution of the sugar raises a third problem that may be called the natural extension of what has gone before, namely the problem of atomism, and with it the problem of the 'grouping' of physical operations. In particular, when the child grasps the conservation of the sugar despite its apparent disappearance, does he believe that the sugar fuses with the water, or does he rather think that it turns into invisible but discontinuous corpuscles? Does he adopt Bachelard's 'metaphysics of dust', and, if so, can we still maintain that the idea of conservation implies a reversible composition of operations?

§1. *Method and general results*

The child is shown two glass vessels filled to the three-quarter mark with water. The vessels are placed on a balance to demonstrate that they are of identical weight, and the child is asked what will happen if a lump of sugar is dropped into the first glass. He may say that the water will rise for one reason or another, or else he may make some irrelevant remark. The original level is marked with ink or an elastic band, two or three lumps of sugar are dropped into the vessel and the new level is marked. Next, the child is asked to predict the level of the water once the sugar has dissolved. He is also asked to weigh the glass of water before the sugar has dissolved and to predict its weight after dissolution. While the sugar dissolves he is asked if it stays in the glass and if so, in what form? Will it be as pure as tap water? What taste if any, will the water have, will the taste persist, and if so, why? Once the sugar has dissolved, he is finally asked to determine the level and the weight of the solution and, having done so, he is asked why the level did not drop back, whether the weight has remained constant, and what has happened to the sugar.

Now, interestingly enough, when these questions were put to more than a hundred four- to twelve-year-olds, we discovered that their responses could be fitted into precisely the same stages as occurred with the clay ball. Thus during Stage I neither the volume nor the weight are conserved, and the substance is thought to vanish completely; during Stage II the substance is conserved but neither the weight nor the volume; Stage III sees the construction of the conservation of weight; and Stage IV that of the conservation of volume. The existence of these stages is proved not only by the average ages with which each is associated—a questionable argument—but also by the fact that the grasp of the conservation of weight invariably goes hand in hand with a grasp of the conservation of substance, and that the grasp of the conservation of volume always goes hand in hand with a grasp of the conservation of weight, but not *vice versa*. Moreover, during Stages II–IV, an increasing number of subjects moves very gradually towards atomism, and it is this advance which we shall be examining in Chapters 4–6.[1]

§2. *The first stage: non-conservation of substance,*
weight and volume, or complete disappearance of the sugar.
(1) *The child's spontaneous attitude to the data*

Stage I, like the corresponding stage described in Chapter 1, is characterized by total reliance on immediate experience, and hence by a refusal to reason deductively. Thus, as soon as the sugar has disappeared from sight, the child concludes that it has been destroyed. No explanation is given of this vanishing trick and no amount of contradictory evidence (e.g. that the water level does not drop back after dissolution or that the weight remains unchanged) suffices to shake this conviction.

The only permanent effect children at this stage attach to the sugar is that it makes the water taste sweet, though only for a time. Most consider the sweetness an extraneous factor, in no way connected with the conservation of matter, weight or of volume.

Before looking at their responses, we should mention that it is often difficult to decide whether they really imply the non-conservation of substance, i.e. whether these subjects really believe the sugar has ceased to exist or whether they think that, as soon as it has been dissolved, it loses its weight and volume and continues as some kind of formless matter. Whenever this difficulty is not due to verbal confusion — in which case the method of interrogation has to be improved — but to the subject's own thought processes, we must pay special heed to it, because, in that case, it reflects the true problem the child faces in constructing the concept of durable substances.

Here, now, are some typical Stage I reactions:

JEJA (6; 1). 'What will happen if I drop these lumps of sugar into the water? —*They will make sugar water.*—Will the water rise?—*No, it won't, because the sugar is light. It just turns into tiny little things.*' Three lumps of sugar are dropped into the water. 'Take a good look!—*Oh, the water has risen.*— Why?—*I don't know.*—What's going to happen next?—*The sugar will melt.*—What does that mean?—*That there won't be any left.*' Once the sugar has dissolved, the demonstrator continues: 'What's it taste like now?— *Sugary.*—Why?—*When the sugar melts it gives its taste to the water.*—So is there still some sugar in the water?—*No, the sugar isn't there any more.*—Will

[1] It might be objected that our experimental criterion — the constant level of the water — is invalid because the volume decreases due to several extraneous factors: part of the energy is dissipated as heat; liquid sugar is slightly less voluminous than crystal sugar; the volume of the molecules changes slightly with changes in temperature, etc. However, all these changes do not affect the issue, for we are not so much concerned to discover how our subjects interpret the phenomena in detail. All we want to know is whether or not they grasp the idea of (absolute or partial) conservation during a physical transformation.

the water still be sugary in a few days' time?—*Yes . . . no.*—Yes or no?—*No.'*

A little later: 'Where has the sugar gone now?—*It's disappeared.*—Look at the water, did it stay up?—*Yes, once it rises it stays up.*—Why?—*I don't know.*—Will you weigh the glass again?' He weighs it. *'The weight is as it was.*—Why?—*It's heavier because there's more water in it.*—Why is there more water in it?—. . .—Where does the extra water come from?—*I don't know, from Geneva.*—But why is there more water now?—*It just happened all of a sudden.'*

MAN (6; 4). 'Look at this glass of water. I am going to drop three lumps of sugar in. Will the water stay where it is?—*The water will rise.'* Experiment: 'Yes, a little bit.—And afterwards?—*The sugar will melt.*—What will happen when it melts?—*The water will drop back because the sugar won't be big any longer. You have to put in something big to make the water rise, but now all that's left is tiny little crumbs* (= the few undissolved grains). *Now it's all melted.*—What's happened to the sugar?—*It's melted, there is none left, the water has melted it all; you can't see anything, there's nothing left.*—But where has the sugar gone?—*It has melted in the glass, there's none left; it's gone to the bottom, and then . . .*—Well, what happened then? Has all the sugar gone or is some of it still there?—*It's still there, but it's melted.*—If you drank this water would it have a taste?—*It would taste of sugar.*—What is taste?— *It's like smell, you can smell it but you can't see it.*—So has some of the sugar stayed behind?—*No, all the sugar has melted.*—Has the water dropped back? —*It has stayed where it was.*—Why?—(Reflects.) *It's this ink spot* (the mark) *which keeps the water up.'* The ink spot is removed. *'No, it still doesn't drop back . . .* (Reflects.) *I can't understand it.*—But when the sugar has melted does some of it stay behind or nothing at all?—*Nothing at all.*—Then why didn't the water drop back?—(Reflects.)—When we can't see any, is some of the sugar left all the same?—*None at all.'*

FER (6; 8). 'What taste will the water have?—*It'll be sugary.*—What is taste? —*It's like steam. After a few days it'll all be gone.*—But where does it go to?— *I don't know.*—And now the sugar is nearly dissolved. Did the water stay where it was?—*No, it dropped back.* (Looks.) *No, it stayed where it was.*— Why?—*I can't tell, there are still some little crumbs of sugar in the water but they don't take up any room.*—Will the water stay up?—*No it's going to drop.'* A moment later: 'Has it dropped?—*No, it's remained where it was; but tomorrow it's going to drop for sure, because there won't be any taste left.'*

MAR (6; 9). *'The sugar will melt and make the water taste sugary.*—And afterwards?—*Afterwards, it will melt; we shan't be able to see it. It will all have melted away.*—What does that mean?—*That it won't be there any more.*— And the water?—*It'll drop back, because the sugar will have melted away.*— Will that make pure water?—*No, the sugar has made it sugary.*—So is the sugar still inside?—*No, it's melted.*—Will the water stay sugary?—*No, it won't.*—What about the weight?—*It will be lighter, because there'll be no sugar left once it's all melted.'* A moment later: 'Has it dropped back?— *Yes, it has.*—Look at it carefully.—*It's dropping back a bit* (which belies the facts).—And the weight?—(Weighs the glass.) *It's stayed the same!*— Why?—. . .—If we boiled the water away would there be some sugar left?— *No it's all gone.'*

Here, now, are a few reactions which, though leading to the same final conclusions as the last, nevertheless betray some inkling of the subsistence of 'formless matter'. However, these responses are so episodic that they cannot be treated as characteristic of a more advanced stage, the less so as they are discarded in the further course of the test, or when the child is confronted with the experimental facts (level and weight). At most, therefore, we can speak of a Sub-stage IB, preparing the way for the intermediate reactions of Stage IIA:

ULD (6; 10). *'It's going to rise a little.—*Why?*—It's like when you put your hand in the water; it rises, your hand pushes it up and the sugar does the same.—*And when the sugar has melted?*—The water will drop back. It's going to turn into little crumbs like dust.—*But why is the water going to drop back?*—Because none of it can be seen.—*But you just told me the sugar turns into little crumbs.*—Yes, but after that it'll disappear.—*Will it be gone?*—It'll stay, but it will melt and then we'll have nothing but water.—*Pure water?*—It'll be sugary, because it had some sugar in it.—*And will some sugar stay behind, or will anything take its place?*—No, nothing will stay behind, but it'll be sugary all the same.—*And in a few days' time?*—In a few days' time we shan't see anything.—*Will it be like pure water?*—No, not pure...Oh, yes it will be, if you leave it long enough.'*

Experiment: 'Has the water-level dropped?*—It's stayed because all the sugar hasn't melted, but it'll drop after a few days.—*And the weight?*—Like before, because it's melted.—*Weigh it for yourself.—*(Stupefaction. Changes the two scale pans round to check, then weighs the vessel in his hand.) *I can't explain it.'*

CLA (7; 0). *'It'll make the water rise, because the sugar has some weight and that lifts the water up.'* A moment later: *'The water will drop because the sugar will have melted. We shan't be able to see any.—*Will the sugar still be inside?*—It'll always be inside, but you can't see it.—*And the weight?*—It will be lighter.'* Experiment: 'Has the water dropped back?*—No.—*But the sugar has melted?*—Yes. But there's nothing to pull it down.—*What taste has the water now?*—Sugary, because there's sugar inside.—*And later?*—It won't be sugary because the taste will have gone.—*What happens to the sugar?*—It is a lump to start with, then it gets ground up and flattened, and then it's gone.—*What about the weight?*—It gets less.* (Weighs.) *Oh, no it hasn't.—*Why not?*—There's more water than before.—*Where does the extra water come from?—...I don't know.'*

Such are the main reactions produced by children at Stage I. In interpreting them, we must be careful to distinguish their expectations before the experiment from the responses produced after the demonstration that the weight and level of the water do not change upon the dissolution of the sugar.

The expectations seem perfectly clear: there will be no conservation of volume, weight, or even of the dissolved substance, which is believed to disappear completely. This may be inferred from the child's use of the term 'melted', by which he means that 'there won't be any sugar left' (Jeja), that 'it won't be there any more' (Mar), or that 'it will have gone' (Or, 7; 0). But are we really entitled to conclude from these remarks that the child believes in the total destruction of the sugar, or should we rather take it that he is simply trying to indicate that the sugar is absorbed by the water, that it disappears as a solid but nevertheless becomes mysteriously combined with the water or the air?

There are several signs that might seem to support the second interpretation. First of all, some of the children think that, before it disappears, the sugar turns into 'crumbs' (Man, Fer, *et al.*) or into 'dust' (Uld), which suggests that they may have been thinking of invisible particles lacking dimensions or weight. However, to children at this stage, these particles are nothing but the visible remains of the sugar while it is still in the process of dissolution, not permanent entities in the form of non-perceptible corpuscles. Again, when these children are asked, immediately after the dissolution of the sugar, if it has disappeared and where it has gone, we again obtain responses that seem to corroborate the second interpretation. Thus Man, having first declared that 'there's none left' nevertheless went on to say that 'it's still there, but it's melted'. Similarly Uld seemed to think that the sugar is changed to water, but later added that 'nothing will stay behind' and that even the sugary taste would disappear. Cla similarly explained that the sugar 'will always be inside but you can't see it', and that the water was sugary 'because there's sugar inside' but added that 'the taste will have gone'.

Nevertheless, we do not think that the second interpretation holds for children at Stage I. Moreover, even if it did, it would in no way entail a grasp of the conservation of matter: all our subjects are convinced that the sugary quality decreases upon dissolution. Nor is that all. When we come to Stage II, we shall see that the child attributes the apparent disappearance of the sugar to a kind of transmutation. Younger children, by contrast, merely take cognizance of the disappearance without bothering about the reasons. Hence several factors speak in favour of the first interpretation, however odd it may appear to be.

In the first place, all the answers we obtained were contradicted by subsequent answers. Now, it is true that children are quite used to such contradictions, especially when vacillating between two hypotheses, but the fact that the contradictions appear in such rapid succession suggests that we must be careful not to mistake mere words for genuine opinions.

Second, the hypothesis that the child believes in the total

disappearance of the sugar has greater psychological plausibility, because children at Stage I show no interest in the problem we pose to them, i.e. in the persistence of matter. To these subjects, the dissolution of the sugar involves two completely unrelated phases: before the disappearance the sugar is in lumps which gradually crumble; and then, quite suddenly nothing is left. It is the demonstrator alone who urges the child to guess what has become of the dissolved substance; the child himself neither cares nor wonders about something as obvious as the disappearance.

This is also borne out by his view of what happens to the taste. If any one observation could convince him that the sugar does not disappear, it would surely be the persistence of its sweetness. Now, we find that most of our subjects believe the precise opposite. Fer, for instance, declared quite spontaneously: 'It's like steam, after a few days it'll all be gone'; and even went so far as to declare, after he observed that the level of the water did not drop back: 'Tomorrow it's going to drop for sure, because there won't be any taste left.' Uld and Cla were of the same opinion. Moreover, these very children (except for Fer who was briefly discountenanced when he discovered the constancy of the level), and also those who seemed to assume the permanence of the sugary taste, in no way grasped the connection between the sugary taste and the conservation of matter, a link that seems quite obvious to children at later stages. Jeja, for example, said that 'when the sugar melts it gives its taste to the water', but when he was asked if that meant some sugar was left in the water, he replied; 'No, the sugar isn't there any more.' Similarly Man, who explained that taste is like smell ('You can smell it but you can't see it'), added that none of the sugar was left. When Fer said that taste was 'like steam', he was thinking of its insubstantiality, because he used this comparison to prove that none of the taste would persist. Uld found a formula of his own to express the non-substantiality of taste: 'Nothing will stay behind, but it'll be sugary all the same.' In short, all these children believe that though the taste of the sugar lingers on for a time, it eventually disappears. As one of them put it: 'There is no more sugar left because all that stays behind is the taste, and that will evaporate.' Taste is a quality without a material support, and hence like the shadow that follows an object at a distance and then disappears with it.

All this confirms our view that children at Stage I believe in the total destruction of the dissolved sugar. But, let us repeat, even should they merely believe in a significant reduction of the sugar substance, they still have not the least inkling of the conservation of matter, so that this stage is in clear correspondence with that described in Chapter 1. This conclusion applies *a fortiori* to the conservation of weight, as witness their predictions about the weight of the two vessels, before and after the dissolution of the sugar in one of them. All our subjects realize that if two vessels (A) and (B) contain the same quantity of water and are of the same weight, then (B) must become heavier when three lumps of sugar are added to it. But when it comes to predicting the relative weight of (A) and (B) after the complete dissolution of the sugar in (B), all of them contend that the weight of (B) will become equal to that of (A), i.e. that the sugar loses its entire weight upon dissolution. As Mar put it, 'It will be lighter, because there'll be no sugar left once it's all melted.' Cla, for his part, declared that the weight 'gets less' once the sugar has dissolved, and Uld explained that, before the sugar dissolves, 'the sugar and the water are heavier because there is sugar in', but he failed to conclude that the weight of the sugar must be conserved even after dissolution. In brief, all these subjects believe that the weight diminishes as the solid fractions disintegrate. Now, at this stage, weight, volume and substance are still considered inseparable, so that the child freely associates the non-conservation of weight with that of substance or, conversely, the non-conservation of substance with that of weight.

Needless to say, he reacts similarly to the conservation of volume. All subjects at this stage, having seen that the water level rises when the three lumps of sugar are immersed, expect that the level will drop back to the initial mark once all the sugar has dissolved. Since they realize that the sugar takes up room in the water, they clearly believe that the apparent destruction of the substance leads to the total disappearance of the volume. In this connection, we can do no better than look at the child's reaction when he discovers (it matters little whether or not he anticipated it) that the water level rises upon the immersion of the three lumps, for it is at this point that his failure to dissociate volume, weight and substance first becomes plain. Most of our subjects mentioned the weight of the lumps. Jeja, for example, did not think that the water would rise, 'because the sugar is light'. Those children who predicted that the level would rise used the same argument in reverse: 'It'll make the water rise, because the sugar has some weight and that lifts the water up' (Cla). But in their view, weight is a kind of force or the cause of an upward current, proportional to the volume (see also Chapter 7). This explains why Man used the term 'big' to refer to both the weight and the volume, and why Uld compared the immersion of the sugar to the immersion of one's hand: 'Your hand pushes it up and the sugar does the same.' In short, the water rises because the sugar lumps are both strong and 'big', the second term referring to their combined weight and volume. Now, except for Jeja, according to whom the level would not drop back after dissolution, for the purely phenomenal reason that 'once it rises, it stays up' (he was the only subject to predict that the level would *not* rise), all these children assumed that, as the sugar dissolves, it loses its volume, weight and force, whereupon the water returns to its original level: 'The water will drop back because the sugar won't be big any longer' (Man).

In other words, before he is brought face to face with the experimental facts, the child at this stage carries the idea of non-conservation to the point of total destruction, so much so that he will even assert the non-substantiality of the sugary taste. Here we have further evidence of the egocentrism and phenomenalism characteristic of this stage: what is no longer perceptible no longer exists, and all things are what they appear to be on direct inspection.

§3. *Stage I: the complete disappearance of the sugar.*
(2) *Reactions to the unexpected experimental data*

In the second part of the questionnaire, the child is presented with two crucial facts that his elders would consider certain proof of conservation, namely: (1) that the vessel containing the dissolved sugar weighs as much as it did immediately upon immersion of the lumps; and (2) that the water level has remained constant. How do subjects at Stage I react to data that so plainly belie their predictions and spontaneous interpretations?

Now — and only empiricists and uncritical realists will be astonished by this fact — our subjects close their eyes the more resolutely to the experimental results the more enmeshed they are in phenomenalism. The methodical rule followed by children at Stage I seems to be that nothing is true that cannot be the object of sense experience. This explains why they cannot grasp the idea of conservation which calls for a mental construction apparently in conflict with the data of visual perception, and why, having discovered that the weight and the level of the water have remained constant, they are not in the least perturbed. The apparent paradox disappears as soon as we recognize that all real experience involves precisely this kind of intellectual construction and deduction and that the two are not contraries except to egocentric phenomenalism.

How, then, do our subjects react when they see that the water level remains constant after the sugar has dissolved? The most cautious among them claim that they are at a loss. 'I don't know,' said Jeja; 'I can't understand it', Man confessed; while Mar and others sighed and fell silent. Others again simply deny the facts: 'It dropped back,' said Fer without looking, and Mar was of the same opinion. Alternatively, they predict that the level will drop later: 'It's remained where it was,' said Fer, 'but tomorrow it's going to drop for sure because there won't be any taste left.' 'It'll drop after a few days,' added Uld. A third type of reaction involves attempts to produce an explanation but of a most peculiar type: various data of immediate experience are linked together at random. That is why Jeja explained that 'once it rises, it stays up', or 'it just happened all of a sudden', which amounts to fitting the facts into a kind of legal, non-causal, relationship. Cla went further still: he at first assumed that the water would rise upon the immersion because of the weight of the sugar, and even went so far as to say that 'it'll always be inside but you can't see it', and also granted that the water will 'drop because the sugar will have melted'. However, when his second prediction was proved wrong, he did not attribute this strange discovery to the possible persistence of the sugar; as a convinced phenomenalist and disdaining explicative deductions, he contented himself with saying that the water did not drop because 'there is nothing to pull it down'. There could be no better way of marking his determination to stick to the immediate data: the sugar has melted, 'the taste will have gone', but the level stays up because the lumps have caused it to rise and no new cause has intervened 'to pull it down'! Finally Man, who ventured to make an attempt at correlation, assumed that it was 'this ink spot' (= the mark on the glass) which 'keeps the water up', thus providing a striking example of a purely phenomenalistic explanation.

Much the same is true of the way in which these children interpret the experimental discovery that the weight remains constant despite their predictions that it will decrease. From this discovery, they never conclude that the sugar must have remained in the water. 'It's heavier because there's more water in it,' Jeja explained, adding that the surplus came from the tap like the rest, i.e. 'from Geneva'. When pressed further, he invoked what we might call the *ultima ratio* of phenomenalism: 'It just happened all of a sudden.' Uld was more cautious: 'I can't explain it,' he admitted; similarly Cla said, after some hesitation, that he did not know where the extra water came from.

These reactions are of great importance because they reveal, better than anything else, the characteristic conflict children at Stage I experience as a result of the incompatibility of their phenomenalist system of global qualities, juxtaposed or fused but never composed, with the deductive system of co-ordinated relations that will eventually lead them to conservation and atomism. During the entire test, the child's attention is riveted to a number of intuitive data which he examines in turn: the appearance of the sugar, in lumps and then in grains, and its eventual disappearance; the taste of the water after dissolution; the height of the level and its constancy; the weight of the sugar water before and after dissolution, etc. To interpret the two final experiments correctly he must correlate the various qualitative data, which calls for an operational schema that is both logical and quantifying and which alone enables him to proceed from one discovery to the next without becoming bogged down in contradictions. Now, at Stage I the child has realized perfectly well that the sugar causes the water level to rise, but he fails to co-ordinate this perceptive datum with the fact that the water does not drop back when the sugar has 'crumbled' or when it has completely dissolved. Again, he has discovered that the water with the three undissolved lumps weighs more than the pure water, but he in no way links this observation to the fact that the weight remains constant even after dissolution. He also fails to associate the persistence of the sugary taste with either the volume of the water or with its weight. The appearance of 'crumbs' does not strike him as the first step in a process of segmentation that continues beyond the limits of perception: once the 'crumbs' have disappeared 'nothing at all is left'. In short, everything happens as if the child merely registers the data he perceives but does not fit them into a coherent whole.

Admittedly, no quality is ever perceived in isolation, and even while he is still confined to egocentric phenomenalism, the child establishes elementary relationships, thanks not only to his empirical discoveries (for example, that the water level rises when a lump of sugar is immersed), but also thanks to the pre-relations suggested by his own actions (for example: the sugar pushes the water up as his hand would: 'Your hand pushes it up and the sugar does the same'). Nothing is entirely passive in psychic life, and perception paves the way for what relations operations will eventually transform and complete: an operation is both the continuation and the correction of intuition. But these primitive relations are still confined to the sphere of *actual* perception, e.g. to the relationship between the dissolution of the sugar and the taste of the sugar water, or between the rise in level and the increase in the weight of the water. To construct an overall system, by contrast, the child must first treat these actual states as so many results of physical operations, i.e. of *reversible* displacements in time and space. Now, it is precisely this kind of reversible construction that still eludes children at Stage I: even when they have established the actual relations, they continue to juxtapose successive states instead of linking them operationally.

For example, once perception has shown the child that the water level rises upon the immersion of the sugar, he might easily conclude that (1) = 'the sugar takes the place of the water', whence (2) = 'the displaced water occupies the space above the initial level'. Once he has constructed this system, a reversal of the operation (1), i.e. the removal of the sugar from the water, would lead him to (1A) = 'the empty space left by the removal of the sugar is filled by the water', whence (2A) = 'the water returns to its initial place, and evacuates the space above the original level'. But does the child really argue in this way when he predicts that the level will drop after the dissolution of the sugar? He begins by postulating (3) = 'the dissolved sugar ceases to exist', which would seem to corroborate (1A) and (2A). However, this construction is not a reversible transformation, and proposition (3) prevents the inversion of the operation (1A), i.e. the composition of the relations is no longer possible because one of the terms has been lost on the way. Moreover, when the child discovers that the level has stayed up, he must, in order to reconcile (3) with (2A), deny (2A), and postulate (4) = 'Once it rises, it stays up' (Jeja's hypothesis) or 'There's nothing to pull it down' (Cla); next, in order to reconcile (4) with (1A) and with the constancy of the weight, he must invoke the appearance of more water, or (5) = 'It just happened all of a sudden' (Jeja), which, as it were, introduces a creation *ex nihilo* to offset the destruction (3) of the sugar. We could similarly demonstrate the incoherence and irreversibility of the relationship between dissolution and weight or taste. In short, the child 'translates' (Stern) or 'postduces' rather than 'deduces'; i.e. he constantly changes his system of composition by fusion and juxtaposition and not by reversible co-ordination.

Now this irreversible and pre-operational character of the relations perceived or established by the child not only explains the general failure to grasp the idea of conservation at Stage I, but it also, and above all, makes clear why the child does not advance from the changes he has observed, e.g. from the 'crumbling' of the lumps into ever-smaller particles, to the idea of atomism, and why at subsequent stages, when he adopts an operational approach, the same observations lead him to conclude that it is possible to decompose as well as to recompose the lump. G. Bachelard has shown[1] that the intuitive models on which nascent atomism is based are powder and dust. Now, sugar, which our subjects can observe daily both in powder form and also in lumps that break

[1] G. Bachelard, *Les Intuitions atomistiques*, Paris, Boivin, 1933.

up into 'small crumbs', would seem especially favourable to the construction of this type of model. But why does the 'pulverization' of the lump not lead them directly to atomistic explanations? This is what we shall now examine in brief.

All children at Stage I notice the 'pulverization' of the sugar during its dissolution. Thus Jeja contended that the sugar 'turns into tiny little things' and that, as soon as it had dissolved, 'there won't be any left'. Similarly, Man thought that the level would drop because 'all that's left is tiny little crumbs', but added immediately afterwards, 'It's gone to the bottom and then ...' — eloquent silence. Fer added that 'there are still some little crumbs of sugar in the water but they don't take up any room'. Finally, Uld stated explicitly: 'It's going to turn into little crumbs like dust', and then added, 'but after that it'll disappear.' In short, all these reactions are so many elements of a future 'powder metaphysics', but still so bound up with phenomenalistic perception that they do not culminate in true atomism, i.e. in a form of atomism based on mental constructs rather than on pure perception. Now, nothing could be simpler than this final step: all the child has to do is to realize that the 'tiny things', the 'crumbs' or the 'dust', result from the break-up of the initial lump and that, if this process is continued, the end product will be a host of invisible but none the less substantial corpuscles. The best proof that this step is easy to take is that so many children take it at Stage II. Why, then, do children at Stage I fail to adopt a hypothesis that would prove so useful to them in escaping from the contradictions in which phenomenalism has enmeshed them?

The reason is quite simply that they do not conceive of the disintegration of the lump in an operational manner, but that they treat it as a spontaneous process: the sugar 'just turns into tiny little things', as Jeja put it. It goes without saying that if the disintegration is conceived as a spontaneous process it cannot be considered reversible, so that it is only logical to think the sugar is destroyed and cannot be reconstituted. But as soon as that process is treated as an operation then, however small and invisible the atomistic fractions become, the child will always be able to produce a mental inversion of the operation and to deduce from the resulting reversible composition that the substance of the sugar is conserved, no less than its weight and volume.

We may therefore take it that the various elements of all these Stage I responses are perfectly coherent. In their spontaneous expectations before the experiments, no less than in their reactions to the latter, all these children evince a clear incapacity for operational construction, and hence fail to arrive at groupings capable of lending support to conservation or to atomism. More-over, they believe that the substance disappears in much the same way as children at the sensori-motor phase of mental development (before the end of the first year of life) believe that objects are destroyed when they disappear from view. But though they thus prove phenomenalists to the point of remaining deaf to the experimental findings (or to rational experience), their phenomenalism cannot be divorced from egocentrism, which assimilates the perceptive data to action schemata (in which the weight-force, the taste-quality, and the 'crumbs' are due to a biomorphic process, and the substance is reduced to a waxing or waning force). The two extreme forms of intellectual activity thus stand revealed once again as phenomenalistic egocentrism and the grouping of reversible compositions.

CHAPTER FIVE

The Conservation of the Sugar and the Beginning of Atomism

At Stage I the substance of the sugar, no less than that of the clay (see Chapter 1, §2), is barely distinguished from the qualities it supports: it is conceived as a dynamic force comparable to a vital principle, and hence as being subject to growth and destruction. It is $\phi\acute{v}\sigma\iota\varsigma$ qua vital energy, not the primary and constant substance of the pre-Socratics.[1] Stage II, by contrast, which we shall now examine (and which corresponds to the second stage described in Chapter 1, §3 and §4), sees the conservation of substance, though not yet the quantification of weight or volume. This elementary form of conservation is therefore based on the persistence of matter and involves both intuitive transformations and also the quantification of matter as such. The transformations fluctuate between a sort of evolution or metamorphosis (the sugar becomes changed into water), and a kind of atomistic composition (the lump of sugar splits up into invisible grains devoid of weight or volume), with all sorts of intermediate attitudes (the grains are changed into water, etc.). The quantification is the same in both cases, though it is more strongly reinforced by the child's nascent atomism.

§1. *The first sub-stage of the second stage (Stage IIA):*
intermediate reactions

Between those children who believe in the total destruction of the sugar (see Chapter 4) and those who come down unequivocally in favour of its conservation, there are a large number of intermediate cases which merit closer attention. These subjects are still more or less convinced that the dissolved sugar is destroyed, but feel compelled, when observing that the water level and the original weight remain constant, to seek an explanation in terms of the conservation of the sugar, though they still hesitate and contradict themselves during successive observations:

[1] C.A. Berger, *La racine* $\phi v\omega$, Paris, Champion, 1925.

GRI (6; 10) begins with a typical Stage I reaction: 'What will happen when we put the sugar in the water?—*We shall see it for a bit, and then we shan't see it any more. It will melt and won't be in the water any longer.*—Where else will it be?—*Small bits of it will remain, then it will all be like castor sugar, and then the lot will have melted away. None of it will stay; the water will be as it was.*—What will it taste of?—*Of sugar.*—And later?—*It won't be sugary any more, because all the sugar will have melted away and there won't be any left.'* Level: *'It has risen because the sugar takes up room.*—And when all of it will have melted?—*It'll drop because there won't be any sugar left in the bottom.'* Weight: *'It'll get lighter again, because there won't be any sugar left.'*

But after observing that the weight does not change, Gri changes his mind: *'Some of the sugar has stayed behind all the same.*—But why is the weight the same when it's all melted?—*Because there is always some sugar left, but as a powder so we can't see it.*—Look where the water is now!—*It hasn't dropped back, because there is still a little bit of powdered sugar we can't see.'*

BUR (8; 4) wavers between conservation and non-conservation: *'All the sugar will melt.*—What do you mean by that?—*I mean that it will grow smaller and then we shan't be able to see it.'* Level: *'It's going to rise because the sugar is heavy* (downward gesture)... *It's dropped hard* (= quickly) *and it's made the water rise.'* On further dissolution Bur thinks the level will rise even more, but it will drop back again: *'At first, the water will keep rising a bit.*—Why?—*Because of the sugar. There is more water than before because of the scraps.*—What scraps?—*When the bits of sugar melt they lose some scraps and these scraps rise to the top.'* But a moment later he claims that the sugar has *'disappeared'.*—Did something take its place?—*No. It disappeared because it melted.*—Did some of it stay?—*Nothing.*—Where did it go to?—*Into the glass.*—But is it still there or not?—*The sugar isn't there any more, but the scraps have stayed in the water. The sugar has gone.*—What's happened to the sugar then?—*It's turned into small grains.*—And what about the scraps? —*They're smaller than the grains.*—Is there anything left in this glass?—*Yes, the scraps of melted sugar.*—Where are they?—*In the water.*—So where is the sugar?—*There isn't any left, it's all melted.*—What does that mean?— *That it's all gone.'*

Next Bur is asked to compare the weight of a glass of pure water plus three lumps of sugar with the weight of a glass in which three lumps of sugar have been dissolved. *'This one* (the second) *isn't as heavy, because all the sugar has gone.*—And before?—*They weighed the same, but now that it's melted the weight is no longer the same, the lumps have gone.*—Did they change into something else?—*Yes, into scraps, into tiny little grains.*—What would happen if we weigh the lumps on one side of the balance, and all the scraps on the other?—*The sugar will be heavier, the scraps are smaller.*—And if we weigh a whole lump of sugar against a crushed lump?—*They will weigh the same; they are still the same scraps.*—And what about these two pans?— *They'll be the same as long as the sugar hasn't melted. But it's different when it's melted, because then all the sugar will have gone. It gets smaller and smaller until nothing at all is left.'* When Bur is shown that the weight and level remain constant, he finally decides in favour of the conservation of substance.

PFI (8; 6) similarly hesitates between the two solutions but ends up in favour of atomism: *'The sugar will disappear; it'll get smaller and smaller. In the end none of it will be left.*—What taste will the water have?—*Sugary.* —And in a few days' time?—*It'll get stale and the taste will have gone. The level will rise because the lumps take up room.*—And when they have melted? —*They'll still take up some room, but less.*—And is the sugar still inside once it has melted?—*Yes, a tiny bit remains.*—What will it look like?—*It will all have melted, we shan't see it any more, it will have disappeared.*—And will the water drop down again?—*Not straightaway.*—Why not?—*The sugar has weight even when it's melted.*—The sugar inside?—*Yes, we can't see it, but it's inside all the same.'* Pfi thus anticipates the partial conservation of the volume, weight and substance simultaneously, and is astonished to find that neither the weight nor the level has changed in any way. He quickly adapts himself to the new facts: *'It hasn't budged.*—Why?— *Because even though we can't see the sugar it still has the same weight.*— What's happened to the sugar?—*It's in crumbs, in tiny little crumbs that nobody can see.*—Not even with a lense?—*No, they're much too small.'*

BAL (8; 7) similarly starts off by saying: *'The sugar will have disappeared, there won't be any left.*—What taste will the water have?—*Oh, I see, the sugar has disappeared but some of it stays behind all the same to make sugar water.*—How so?—*It melts completely, and then there is nothing but sugar water.*—And so?—*The taste stays all the time.'* As for the weight: *'It'll get lighter when the sugar has melted.*—And will the water stay up or will it drop back again?—*It'll stay high because the sugar was heavy when you put it in, it made the water rise, and so the water will stay high.*—Is the sugar still in?—*No, it's not.*—Will the sugar water weigh as much as the other?—*No, because you've put sugar in.*—So has something of the sugar stayed behind? —Yes, its taste.*—Why will this glass be heavier than the other?—*Because the sugar in it has melted, and there was none at all in the other glass.*— So does some of the sugar stay?—*Yes, the taste.*—Does the taste weigh anything? —No, it doesn't.*—Why will it be heavier?—*I can't think.'* The glass is weighed and Bal can see that its weight has not changed: 'Well, has the sugar stayed inside or not?—*No, it hasn't, but it's liquid now; there is some sugar-juice and that has the same weight as the sugar. The juice is runny.'*

GO (8; 11). *'The sugar gets smaller all the time and then it nearly disappears.*— Won't there be any left in the glass?—*No, there won't be, but there will still be its taste because it's melted.*—Does the taste stay on?—*No. Oh, yes, it does.'* Weight: *'When the sugar has melted it'll be the same weight as when there's none.'* Level: *'It'll drop back once it's all melted, because the lumps will have gone.'* But after the experiment, he says: *'Some tiny little grains stayed in.*—And if we put them together, would that make three lumps?—*No, some of it melts and some of it stays put.*—And in the end?—*It won't drop back straightaway; some of the little grains will stay at the bottom.*—They won't have melted?—*Yes, they will melt in the end.*—So why won't the water drop down again?—*Because of the sugar in the water.'* Weighing: *'The glass with the melted sugar is a little heavier than the other one because there is still some sugar inside.'*

NOL (9; 4). *'It turns into small crumbs.*—What taste will the water have?— *Sugary, because the sugar has gone and it's sugared the water. The sugar grains have come up.*—And in a few days' time?—*It won't be sugary any longer.*—Will none of the sugar be left?—*No, none.*—It'll all have gone?—*Yes.*—But how is that possible?—*Ah!* (changes his mind) *the grains can't go away, they can only go as high as the water* (to the surface) *and up there they all melt.*—What does that mean?—*That the grains don't exist. All that remains is a little bit of water. It's like snow when it melts, all that remains is a little bit of water.'* Experimental determination of the level: *'Oh, it's stayed up. That's because the sugar has been turned into juice, so there's a little bit more water now.'* He anticipates that the weight will be *'lighter when the sugar has melted, because there is none left'.* The glass is weighed and he gives the same explanation as he did for the volume.

COL (9; 6) thinks that the volume of the sugar disappears and that *'only the taste stays behind'.* The weight vanishes as well: *'It weighs the same as when there was no sugar in.'* But after watching the level, he explains that *'when the sugar melts it makes some extra water; the sugar gets changed, so you have thick water with the taste of sugar'.*

BAG (9; 8). *'The water will drop down again because it hasn't the extra weight, so it doesn't take up any extra room.*—Why not?—*The lumps have gone; first they get very small and then they're suddenly gone.'* But after seeing that the level remains constant, he says: *'The water stayed on top because all the sugar has changed into sugar-water.'* However, he still anticipates that the weight will disappear: *'It won't be heavy any longer, it's sugary now, but later there'll be nothing left at all, it'll be pure water.'* But seeing that the weight remains constant, he corrects himself: *'The lumps have changed into sugar-water.'*

GIL (9; 10) also starts out by saying that the volume and weight will vanish *'because all the sugar will have dissolved.*—What does that mean?—*All of it will have gone.'* But observing that the level and weight remain constant, he adds: *'I think that the sugar was heavier all the same. It's dissolved, but that makes it heavier all the same because the sugar stays inside even though we can't see it any longer: it's in such tiny pieces that we can't see them.'*

MAT (9; 10). *'It'll be the same weight as the pure water because the sugar will have gone.'* Then, recalling that the water is 'sugary', he goes on to assert that *'the sugar is still there but it's nothing but small grains at the bottom.'* Moreover, *'all the small grains get scattered and that sugars the water.'*

We have dwelt at some length on these responses because of the light they throw on the dawn of conservation and atomism. In fact, all these children, like those at Stage I, are still inclined to think that the material aspects of the sugar vanish in the course of its dissolution. However, instead of adhering to the simple phenomenalism of the younger subjects, they have, in fact, advanced in two respects. Some, like Gri, Bag and Gil, began with a clear denial of conservation, but when the observation of the level and the balance undeceived them they went on to search for an explanation in terms of conservation. Others (the majority) such as Bur, Pfi, Bal, Go, Col and Mat, showed right from the start of the test (or during the test, but in either case before being presented with the experimental data), that they had some grasp of the conservation and tried to reconcile it with the phenomenalist discovery that the sugar has vanished. However, these two advances are, in fact, a single one, and that is precisely why we have lumped the two types of response together. Thus those of our subjects who came to accept the idea of conservation only when forced to do so by the experimental facts (the first group) nevertheless employed new concepts because the same experimental data leave children at Stage I totally unconvinced. This shows that a deductive construction is a *sine qua non* of the correct reading of experimental data. On the other hand, those subjects whom reason alone seems to lead to conservation (the second group) are undoubtedly influenced by past observations or by the questions the investigator puts to them during the test. In either case, therefore, we have both progress in reasoning and progress in heeding the data of real, i.e. constructed, as opposed to immediate, experience.

It should be noted, first of all, that the intermediate responses of both these groups show that the child rarely arrives at the complete conservation of the sugar, i.e. at the idea that the quantity of matter is constant. In most cases, he simply assumes that something persists and is content to fathom in what form; he never bothers to ask himself whether or not the total quantity of matter remains invariant. But this incipient grasp of qualitative permanence undoubtedly represents great progress over the idea of total destruction characteristic of Stage I.

The fundamental discovery that some of the substance persists even after the dissolution of the sugar must, of course, be attributed first of all to experience itself: the persistence of the taste in the 'sugar-water' is the springboard of the new construction. Moreover, the invariance of the weight and the level of the water, no matter whether or not the child discovers them during the final verifications, also play an essential role in this process of discovery. In short, we must not underestimate the part experience plays in the genesis of what is, in fact, the dawn of conservation. It is only when the child comes to realize that the invariance of the entire quantity of matter is an *a priori* logical necessity, i.e. when he makes use of deduction, that we may say he has gone beyond immediate experience; so long as he attributes the conservation to only part of the sugar he continues to rely exclusively on the observable facts. This explains why neither the sugary taste nor even the discovery of the constancy of the weight and the level shake the belief of children at Stage I that the sugar is completely destroyed, while the same facts serve to convince our present subjects that it is conserved. Those who assert the complete disappearance of the sugar know perfectly well that the water is sugary, but they explain this fact by claiming, *inter alia*, that the taste is 'a vapour' unconnected with the sugary substance and bound to disappear in a few days' time. Similarly, if they discover that the water level stays up and that the weight remains constant, they will tell you that some extra water was mysteriously created during the experiment. In short, the perceptive relations supplied by direct experience do not suffice to engender the conservation of substance: as long as the child remains on the plane of egocentric phenomenalism characteristic of Stage I, he cannot subject direct experience to deductive composition, and that is why his observations do not shake his belief in the complete destruction of the sugar. How, then, do our present subjects manage to arrive at the idea of conservation from these very observations, or rather what enables them to resist their natural tendency to believe in the total destruction of the sugar?

In fact, inasmuch as the experimental induction by which the child of this level is carried beyond egocentric phenomenalism is already a construction, it constitutes an incipient composition, and the latter becomes deductive as soon as the system formed by the newly co-ordinated relations culminates in a reversible grouping. This is why Stage IIA is of such great interest: it marks the beginning of a process leading from direct and subjective experience to rationality and operational deduction.

The re-interpretation of the taste, first of all, is highly significant: from being fleeting and non-substantial the sugar flavour has become something durable and the manifestation of a substance. Thus Gri, Pfi and Rol, like subjects at Stage I, all began by assuming that the taste would disappear sooner or later: 'It won't be sugary any more because all the sugar will have melted and there won't be any left' (Gri). In a few days' time 'it'll get stale and the taste will have gone' (Pfi). 'It won't be sugary any longer' (Nol). Bal, Go, Col and Mat, on the contrary, believed that the taste would linger on, and were thus led to the conservation of the sugar. Thus Go thought first of all that the sugar 'nearly disappears', but added 'there will still be its taste'. Asked whether the taste would stay, he replied, 'No. Oh, yes, it does', thus accepting the idea of conservation: soon afterwards, in fact, he contrasted the state (B) 'when the sugar has melted' to the state (A) 'when there was no sugar', thus making it clear that, to his mind, the glass in the state (B) contains something that, though devoid of weight or volume, is nevertheless substantial. Similarly Bal began by saying 'the sugar will have disappeared, there won't be any left', but when he was asked about the taste of the water, he made this telling reply: 'The sugar has disappeared but some of it stays behind all the same to make sugar-water ... the taste stays all the time.' Now it was this very idea that led him to the conservation of the substance (though not of the weight): 'The sugar has gone,' he said, thinking of its weight, but some of its substance remains, i.e. its taste. Col, for his part, stated just as categorically that the volume and the weight vanish while the substance of the sugar—the taste—persists; and Mat was led from the taste straight to the idea of atomism (see below). Other subjects (e.g. Dag), by contrast, did not argue about the taste until after they had arrived at conservation by other methods (weight and volume).

But what happens in the child's mind between the time when he considers the taste a purely momentary, i.e. non-substantial, phenomenon and when he grasps its permanence and hence the persistence of matter? Clearly, immediate experience cannot be responsible for this change, for before the child decides to check whether or not the sugar-water retains its weight after a few days, he must clearly have some inkling of the concept of conservation. Even the idea that a substance is responsible for the taste is a construct, not a percept. In short, the reason why our subjects come to suppose the permanence and substantiality of the sweet taste, is that they have gone beyond egocentric phenomenalism—according to which qualities can exist without substantial foundations simply because they have been perceived—and that they now search for objective co-ordinations. In fact, there is no other solution: either a quality appears to be self-sufficient, in which case it is related unconsciously to the actions of the perceiver, or else it is treated as part of a genuine grouping, in which case it must be provided with a substrate. This is why the two new characteristics of the taste, i.e. its persistence and its substantiality, are one and the same thing: they reflect the beginning of co-ordination decentred from the self and based instead on the reality of physical operations.

The child's approach to the conservation of weight follows a similar course. All our subjects were convinced, before the final experiment, that the weight of the sugar lumps would vanish, or decrease very significantly, upon dissolution. 'It'll get lighter again because there won't be any sugar left,' said Gri. 'Nothing at all is left (of the weight),' said Bur. 'When the sugar is melted it'll be the same weight as when there's none,' Go and others explained. Col, Bag, Gil and Mat were of the same opinion, but Pfi believed that part of the weight would be preserved: 'The sugar has weight even when it's melted'; but not as much as the lumps because the water 'lifted up' by the latter will drop back 'a little' ('not straight away'). Bal, too, said that 'it'll get lighter when the sugar has melted', thus showing that he did not believe in the total disappearance of the weight. Col wavered between the two views. All these children, therefore, believed spontaneously in the destruction or diminution of the weight. Now, as soon as the final experiments made it clear to them that the weight had remained constant, they immediately concluded that the substance had been conserved. Thus Gri, who had previously shown no tendency to think in terms of conservation, exclaimed: 'Some of the sugar has stayed behind all the same', and even added that 'there is always some sugar left'. Similarly Gil who had thought that all the sugar would have gone said after the final experiment: 'I think that the sugar was heavier all the same ... because the sugar stays inside even though we can't see it any longer.' To Bur, Go and Bag, the constancy of the weight merely corroborated their original view (based on taste or volume) that the substance was conserved. Pfi and Bal assumed the conservation of part of the weight from the outset, so that they were not at all surprised by the results of the experiment. In brief, whereas the final weighing fails to persuade subjects at Stage I of the persistence of the sugar, it causes subjects at Stage IIA either to change their minds or else to feel even more certain of the conservation of matter. The new co-ordination has the same cause as the discovery that taste is something substantial, namely that all sensible qualities must have a *substrate* whose persistence and transformations can be fitted into an objective system. Now it is interesting that the co-ordination of weight and substance should not be achieved until the constancy of the weight has been recognized: the child will readily grant that the sugar is conserved in the form of taste or of a weightless substance, but he cannot conceive of any conservation of weight unless he has first grasped the invariance of the substance. Thus Bal, who explained that the water will 'get lighter when the sugar has melted', nevertheless went on to say that the sugar-water would be a little heavier than the pure water 'because the sugar was heavy when you put it in' – an insoluble problem until the constancy of the weight is recognized.

The co-ordination of volume and substance proceeds along similar lines, despite the residual phenomenalism we detect in some of the responses (e.g. of Bur and Bal). Most of these subjects believe that the water level will return to the initial mark once the sugar has dissolved, but as soon as they discover that it does not, they conclude that the sugar has been conserved. Thus Go said first of all, 'It'll drop back once it's all melted because the lumps will have gone', but seeing that the level stayed up he corrected himself and said, 'Some tiny little grains stayed in.' Those who had assumed the conservation of the sugar from the start, for whatever reasons, merely felt confirmed in their views by the experiment. We may therefore say of the volume, as of the weight, that it does not follow automatically from the conservation of the substance, whereas the constancy of the level (and of the weight) leads the child straight to the invariance of the substance. Now the fact that Bur and Bal apparently grasped the constancy of the volume before that of the substance might suggest that this interpretation is false, but all they really did was to assert that the level would stay up for purely phenomenalist reasons: the force with which the lumps fell through the water, the rising of the 'scraps' (Bur), or the 'heaviness' of the lumps (Bal). In other words, they react in much the same way as children at Stage I, according to whom the level stays up 'because there's nothing to pull it down'. What we have here, therefore, is not a precocious grasp of the conservation: thus it was only when he discovered the constancy of the weight that Bal assumed that the sugar is converted into a liquid, and hence arrived at a correct interpretation of the permanence of the level.

In short, both the child's spontaneous co-ordination of the sugary taste with the persisting substance before the final experiment, and also his final co-ordination of the constant weight and constant volume revealed by the experiment with the (deduced) conservation of the sugar—the two new steps taken by subjects at Stage IIA—were due to his advance from egocentric phenomenalism (Stage I) to operational composition (Stage IIB). To egocentric pheno-menalism, as we saw in Chapter 4, immediate qualities are both unco-ordinated and relatively undifferentiated: they are undiffer-entiated when they are perceived simultaneously and combined into one and the same subjective schema; and they are unco-ordinated when they are perceived successively and merely juxtaposed. As a result, volume and weight are confused in a dynamic schema that explains why the lumps cause the water to rise, while the taste of the sugar-water remains quite unrelated to any other charac-teristic of the sugar. Similarly, the discovery of the constancy of the level and the weight does not entail any kind of deductive composition. Children at Stage IIA, by contrast, apply a process of differentiation and co-ordination complementing their percep-tion of qualities or relations, and this incipient composition suffices to explain their nascent grasp of the conservation of substance. In particular, it involves the gradual replacement of intuition by a system of partitions and displacements of the parts in space and time, i.e. of fluent and subjective qualities by mobile objects that remain constant under displacement. Children at this stage, in fact, employ three schemata. The simplest of these is the transmuta-tion of the sugar into water; the most complex, the atomistic pulverization of the lump; and between these two we have pulverization with subsequent liquefaction of the now invisible grains.

The first type of explanation is the most primitive because it simply involves the liquefaction of the sugar and thus barely extends the data of perception. But while it fails to turn the dissolution into a spatial operation and continues to be based on an intuitive and irreversible process, it nevertheless enables the child to consider the sugar, once it has been liquefied, as a constant object whose displacement in the water no longer alters its properties. Col and Bal produced excellent examples of this approach. Both started out with the idea that only the taste persists, not the weight or the volume. (Bal admittedly claimed that part of the weight was conserved, but as some sort of residual pressure that had no direct links with the substance.) However, as soon as they discovered the constancy of the weight and of the level, they immediately dropped their original idea of an imponderable and non-voluminous substance, to conclude that 'the sugar gets changed, so you have thick water with the taste of sugar' (Col), or that some 'sugar juice' is added to the water which 'has the same weight as the sugar' (Bal). In this way, the taste, the weight and the volume are combined into a single invariant, a fact which was expressed clearly by Bag when he said: 'The lumps have changed into sugar-water.'

For another group of these children the conservation of matter is also based on liquefaction, but consequent upon a sort of pulverization that heralds the dawn of atomism. Thus Nol thought that the sugar first turns into 'crumbs' or 'grains' which spread through the water and produce its sugary taste. He was clearly thinking of the tiny particles he could still see moving about before complete dissolution, and which, he believed, must persist throughout the liquid as long as the taste lingers on. However, he also believed that nothing at all would remain in the end. Now what is so interesting to the psychologist is that Nol nevertheless progressed from the idea of irreversibility to that of reversible operations, for when he was asked how it was possible that the grains should have disappeared, the reply that he made showed that he had begun to think in terms of spatio-temporal displacements instead of simple intuitive processes: 'Ah,' he said, 'the grains can't go away, they can only go as high as the water' (= up to the surface). But instead of imagining that these 'grains' would keep circulating as such, which would have been an atomistic solution, he tried to reconcile his discovery with the apparent destruction of the grains and hence contended that the grains had melted 'like snow'! When he subsequently discovered that the weight and level remained constant, he found a ready answer: 'The sugar has been turned into juice, so there's a little bit more water now', which recalls the explanations of the first group. Similarly, Go began by assuming that the sugar 'nearly disappears' but that its taste lingers on in the form of a weightless substance mixing with the water but taking up no space. However, when he discovered the invariance of the level and the weight, he concluded at once that the lumps had crumbled into 'tiny little grains'. But instead of sticking to this schema and explaining the constancy of the taste, weight and volume, by the displacement of these grains, he thought that some of them stayed immobile at the bottom of the glass, while others turned into liquid and moved about ('some of it melts and some of it stays put'). Bur vacillated even more strongly between the idea that the sugar breaks up into small pieces or 'scraps' and the belief that it turns into water. In other words, though he adopted an atomistic approach, he was unable to conceive of all the transformations as spatio-temporal partitions and displacements, and hence kept wavering between non-conservation, conservation with liquefaction, and atomistic conservation.

Finally, a third group of subjects concluded directly from the pulverization of the sugar as to its complete destruction, or else adopted an atomistic explanation of its conservation. Partly for spontaneous reasons and partly under the influence of the experimental findings, they thus arrived at a form of causal composition that replaced the intuitive and irreversible transformation of the sugar with the operational partition and displacement of the corpuscles thus engendered. Gri, for example, who began by saying that 'none of it (the melted sugar) will stay; the water will be as it was', concluded from the discovery that the weight had remained unchanged that the melted sugar had turned from a visible powder ('castor sugar') into an invisible one: 'There is always some sugar left, but as a powder so we can't see it.' Similarly Pfi who, at first, believed that 'in the end, none of it will be left', changed his mind and declared that 'a tiny bit' remains, 'we can't see it, but it's inside all the same'. Hence he had only to discover the constancy of the weight to conclude straightaway that the sugar 'is in crumbs, in tiny little crumbs that nobody can see'. Gil, too, at first declared that all the sugar will go, but when he discovered the constancy of the weight and the level, he said that 'the sugar stays inside even though we can't see it any longer; it's in such tiny pieces that we can't see them'. Finally Mat arrived at the same conclusion: he began by saying that 'the sugar is still there but it's nothing but small grains at the bottom'; but in order to explain why the taste had spread through the water, he said: 'All the small grains get scattered and that sugars the water.'

The most advanced subjects at Stage IIA thus account for both the permanence of the sugary substance, and also for the constancy of its weight and volume which they discover thanks to the final experiments, by means of a system of physical operations thanks to which all the data can be co-ordinated: the sugar lump is made up of grains that become dispersed after division of the lump in such a way that each grain is left intact, as is the whole into which all the grains can be re-combined. To the extent that the child succeeds in fitting all these relationships, perceived successively or simultaneously, into a grouping of operations, he comes to grasp the idea of substantial stability and hence to shed the egocentric phenomenalism so characteristic of Stage I. It must nevertheless be stressed, in conclusion, that the incipient atomism of these subjects was only brought out by our deliberate combination of their spontaneous reactions with those evoked by the final experiments. Left to their own devices none of them would have been able to go on from the conservation of the substance, based on the persistence of the sugary taste, to the conservation of the weight and volume. This will become much more obvious in the following sections.

§2. The second sub-stage of the second stage (Stage IIB): conservation of substance but not of weight or volume

After having examined the intermediate cases, i.e. those wavering between phenomenalism and the spontaneous affirmation of the conservation of the sugary substance, we shall now go on to subjects who treat the latter as a logical necessity:

LOU (8; 8). 'The water will be sugary.—And the sugar?—It will melt.—What does that mean?—It turns into small grains, we can't see them but they're in the water all the same.—Are you sure?—Oh, yes, because the water is sugary.—Will it always be sugary?—Yes.—Will the top of the water stay where it is?—It will rise a little bit, it's like putting your hand in a bowl; it takes up room.—And then?—When the sugar has melted, but only when it's all melted, it will come up to the same height as now.'

The demonstrator points to the glass of pure water that serves as a control. 'And the weight?—It's a little bit heavier than pure water. Oh, no, the melted sugar weighs the same as the pure water because when the sugar has melted, the water is more sugary.—And the sugar itself?—It won't have any weight left.—Can't we see it even with a lens?—Yes, we'll still see some tiny little grains.—And do they weigh nothing?—No, they don't.'

Level and weight: 'I would never have thought there would be so much of a difference (between the sugar water and the pure water)! It's the sugar inside; I would never have thought so.—But has the sugar melted?—Yes, but the bits still have a little weight; I can see they have. When it's melted inside it's the *same as if we crumble it all up outside.*—What do you mean by melted?—*Broken into crumbs.*—And under a lens?—*We'd see tiny little crumbs.'*

BON (9; 6). *'The sugar will melt.*—What does that mean?—*We shan't be able to see it; it's in tiny little grains like powder, you can't see those.*—And if we weigh it?—*The melted sugar has no weight.*—And the water?—*The grains take up some room; that makes the water rise and when it's all melted it'll drop back again to where it was before.*—But is the sugar still inside or not?—*Yes, in tiny little bits.*—So why will the water drop back?—*The tiny little bits take up no space.*—But will it weigh more than the pure water?—*No, it won't; the grains are too tiny.*—What are they made of?—*They are made of sugar, in a very fine powder.*—And what about their taste?—*All the tiny little grains keep their taste.'*

SAN (9; 10). *'The sugar will melt.*—What does that mean?—*It'll turn into little balls made of sugar powder.'* The level will rise *'because the sugar takes up room'*, but later *'it will drop back again because the sugar will get small and very fine.'* Weight: *'It'll get lighter again, as before.'* By then *'the sugar will be in tiny little pieces.*—Shall we be able to see them?—*No.*—But then how can we tell that they are inside?—*Because we saw them before.*—But will they still have the same weight?—*No, because they'll be tiny little crumbs, not in one piece.'*

But once presented with the facts (level and weight) San explains: *'it's because of the weight of the sugar that's still inside.'*

HUB (10; 5) begins by explaining that *'melting means the water will get inside; the sugar gets smaller and smaller and the water tastes of sugar.*—Will the sugar still be there?—*It'll mix with the water and turn into tiny little bits like flour.'* Also: *'It will always have the same taste.'* After the experiment, Hub explains that *'there is some water in the sugar, and that makes a little bit more water.'*

OLI (11; 0) believes that when all the sugar has dissolved *'it will have sugared the water and be like dust.*—Will it still be inside?—*We shan't be able to see it any more, but it'll still be inside, all melted.*—And how will it be then?—*It'll be finer than dust and we shan't be able to see it.'* This is why *'the water will always stay sugary.'* The level will drop because *'the sugar takes part of the water for itself, it melts and breaks up.'* Once the sugar has melted *'the water will stay where it was'*, because *'the sugar contains more volume when it's melted'*. Moreover, *'the melted sugar will be heavier than pure water, because the sugar will be full of water.'* But *'once it's all melted it won't have any weight, because the melted sugar weighs less than the lump sugar.'*

From the experiment Oli concludes: *'The volume of the sugar has pressed down on the water and made it rise,'* and the level has *'stayed up, which shows that the sugar has kept its volume; it's still inside but in powder form'*. Weight: *'When the sugar melts it just loses a tiny bit of its weight.'*

JAC (12; 0) thinks that the melted sugar will *'weigh less because it'll be all broken up, like evaporated, melted.*—What do you mean by "melted"?—*In grains, tiny little grains, that get finer all the time.'* The level will drop after dissolution. As for the taste: *'the water will always stay sugary, because the bits at the bottom will give it its taste; the sugar will always stay there.'*

Level and weight: *'It's because of these tiny little balls of sugar that have stayed inside,'* and *'the weight hasn't changed.*—Why not?—*Because none of them have evaporated.'*

The common feature of all these Stage IIB responses is that they reflect a complete grasp of the conservation of substance from the outset, but not of weight or volume. The problem they pose to us is why such firm belief in the conservation of matter should not be extended directly to the weight or the volume.

We need not revert to all the aspects of the co-ordination involved in conservation: these children have merely stabilized the reactions we have described in §1. Not only do they fit all their previous co-ordinations into a single system of spatio-temporal compositions, but it is precisely the limitations or inadequacies of the grouping they adopt which explain why they cannot extend it to the case of the weight or the volume.

Let us recall the three types of explanation examined in the last paragraph: pure liquefaction, pulverization followed by liquefaction, and true atomism.

There is no point in adducing further examples of simple liquefaction, which are moreover rarer at Stage IIB than they are at Sub-stage IIA, and which become increasingly uncommon at subsequent stages, thus showing that this type of explanation is a primitive one. The second type, by contrast, is offered with increasing frequency by children at Stage IIB. Thus Hub claimed that the sugar will turn 'into tiny little bits like flour', but that part of the sugar nevertheless adds to the liquid: 'There is some water in the sugar and that makes a little bit more water.' Interestingly enough, the third schema, i.e. pure or atomistic pulverization, is the one most commonly adopted by these children. Thus Lou spoke of 'tiny little grains' that are invisible, weightless and take up no extra room in the water; Bon spoke of 'tiny little grains like powder' or of 'tiny little bits that take up no space', 'keep their taste', and have no weight; San of 'little balls made of sugar powder'; Oli of weightless 'dust'; and Jac even went so far as to claim that these 'tiny little grains' get 'finer all the time'.

We must now try to determine whether these three types of explanation are so many adventitious representations of the conservation after the event, or whether they reflect the existence of an operational mechanism by means of which the child himself discovers the conservation. Now there is no doubt that the second assumption is closer to the truth than the first, not because the child needs to grasp liquefaction or atomism before he can think in terms of conservation, but because the logical operations needed to construct the conservation are the self-same operations that continually draw him to atomism. How, in fact, do subjects at Stage IIB come to assume *a priori* that the substance of a clay ball is conserved (Chapter 1)? They do so by a double reversible composition of the relations between length, breadth, etc. (displacement of matter) and between the parts and the whole (partition or division of matter), i.e. by two compositions that are either complementary or else fused into a single operation, in which case they lead to extensive quantification. Now we saw in §1 that every attempt at co-ordination by the child, as he progresses from egocentric phenomenalism to the conservation of the sugar, consists precisely in replacing qualitative processes by operations (division and displacement): liquefaction and atomism are therefore the direct results of the operational schemata leading to conservation, the former still participating in the qualitative process, and the latter reconciling the operations of division with those of displacement. Thus when Jac contended that the melted sugar will be 'all broken up, like evaporated' and 'in tiny little grains that get finer all the time', he was still wavering between destruction and conservation, and this because he had failed to reconcile the displacement with the division. However, as soon as he was shown the constancy of the level and the weight, he at once came down squarely in favour of conservation: the 'tiny little balls of sugar' have 'stayed inside' and 'none of them have evaporated'. Clearly, therefore, the child comes to appreciate the conservation of the substance as soon as he realizes that the grains engendered by the division of the lumps are scattered throughout the glass instead of escaping: it is the pulverization and the scattering combined which explain the persistence of the taste and hence that of the substance. In short, the child's nascent atomism constitutes a schema of composition in which egocentric phenomenalism has made way for the complete invariance of matter.

But in that case, why does this composition remain confined to the conservation of matter, and why is it not immediately extended to the weight and the volume? We saw that, prior to the discovery of the experimental data, every one of these children said spontaneously that the atomic grains of the sugar 'have no weight' and 'take up no room'. Now the combined operations of division and displacement ought to have engendered the quantification of the weight and the volume in addition to that of the sugary substance. Why did they fail to do so? What we have here, though in new guise, is the same problem we encountered in connection with the clay balls, and if we return to it here, it is not only because we wish to verify our earlier conclusions but also because this problem impinges on the whole relationship between atomism and the compression or decompression of matter. In the case of the weight and volume of the sugar and also of the clay ball, we notice a shift from egocentric phenomenalism to operational grouping based on actions. Thus if the substance corresponds to the action of retrieving the object, the weight to the action of weighing, and the volume to the action of enclosing or surrounding the substance, the child will quite obviously find that it is easier, during the division of the ball or lump and the dispersion of the fragments, to group actions of the first type than to group actions of the second type, and easier to group actions of the second type than to group actions of the third. This means that phenomenalism and egocentrism persist more strongly with actions of the second type than they do with actions of the first type, etc. In the case of the lump of sugar, which dissolves into invisible fragments and hence does not lend itself to the same mental actions or experiments as the clay, the child finds it fairly easy to imagine that every one of the scattered grains can be recovered. By contrast, the idea of weighing and, *a fortiori*, of measuring the space occupied by something that is 'finer than dust' and quite invisible (Oli) must strike him as quite senseless, as witness Bon's comment that the grains are 'too tiny' to have any weight or to take up any space.

When it comes to the quantification of the substance, weight and volume of the sugar, we thus find that the child has to solve the same problem he encountered with the clay ball, namely to decide whether or not a given object can be divided into pieces and scattered and yet conserve its original qualities. As far as the substance is concerned the answer seems relatively simple: a grain is 'the same' whether we can see it or not, or whether it has been removed from one side or from the other. 'We saw them before,' San said to explain the persistence of the invisible 'balls made of sugar powder', thus indicating that, no matter how scattered they are, the sum of the particles remains equal to the whole, i.e. to the undissolved lump. However, the child finds it exceedingly hard to apply the same composition to the weight. As San pointed out, the sum of the weights of the grains cannot be equal to that of the initial lump, because 'they'll be tiny little crumbs, not in one piece'. Hence Lou's utter surprise when he discovered that the weight had not changed. 'The (melted) bits still have a little weight, I can see they have. When it's melted inside, it's the same as if we crumble it all up outside.' In other words, what prevented these children from arriving at the idea of the conservation of weight spontaneously was a failure to appreciate that the sum of the scattered parts is equal to the whole lump. But as soon as they deduce from the experiments that this sum has, in fact, remained constant, they at once arrive at the correct composition: they treat the 'grains' as parts of the whole, and come to appreciate that the sugar is the same 'inside as out', i.e. that its weight is the same no matter whether it is in a lump or split up into invisible fragments.

Much the same remarks apply to the volume. The child tends to think that whereas the whole lump occupies space, the parts do not, (a) because they are 'very tiny bits', and (b) because they are scattered. The sum of the parts is again unequal to the whole. But as soon as he discovers the constancy of the level, he is led directly to the correct composition: 'It has stayed up which shows that the sugar has kept its volume; it's still inside but in powder form' (Oli); or: 'It's the sugar inside; I would never have thought so!' (Lou).

All in all, the child's reactions to the dissolution of the sugar are therefore fully comparable to his reactions to the deformation of the clay ball. This convergence is the more important as the two situations are quite different: in the present series of tests, unlike the first, the substance is not only deformed but also seems to vanish. Let us now see whether the same convergence can also be observed in the construction of the two other invariants.

CHAPTER SIX

The Conservation of the Weight and Volume of the Dissolved Sugar and the Completion of Atomism

Stage II brought the spontaneous discovery of the conservation of the sugar substance, but not of its weight and volume. At Stage III, the child constructs the second invariant, and at Stage IV he constructs the third. In the course of this development, which we are about to examine, atomism is extended further, and the schema of simple liquefaction is retained.

§1. *The third stage (Sub-stages IIIA and IIIB) :*
the conservation of weight with non-conservation of volume

We shall begin with a few intermediate reactions, i.e. with the responses of children who start out by denying the conservation of weight but arrive at it in the course of the test, either spontaneously or else under the influence of the problems they are asked to solve (Stage IIIA). This distinguishes them from subjects at Stage IIIB who assert the conservation of weight as an *a priori* necessity.

REN (8; 10) assumes that the sugar will stay in the water in the form of *'tiny little crumbs'*, which can *'perhaps'* be seen, but *'only with glasses'*. The level will rise after the immersion, *'because the sugar will take up some space'*, but when all of it has dissolved *'it'll drop back again because the lump will have gone; so the water will take up the same space as before.'* Weight: *'It'll be a little heavier than pure water because the sugar crumbs are still inside.—* Do they weigh as much as the lump?—*No, because when the lump was together it was bigger, and the little grains are all over the water.*—Why will it be heavier than pure water?—*Because of the sugar inside.'* But when a glass of water and a lump of sugar beside it are placed on one pan of the balance, and another glass of water with a dissolved lump is placed on the other pan, Ren predicts that *'it'll be the same weight, because the two glasses are the same, there is one lump to each.*—But will the water drop back or not?—*Yes, because the lump was big and now it's very small.'* He is shown the level marks: *'It hasn't dropped back because the sugar has stayed inside and it takes up room.*—And the weight (experiment)?—*It's just as I said.'*

98

CHA (9; 9). 'Will some of the melted sugar stay behind?—*Some grains will stay at the bottom, like salt or castor sugar.*—And the weight?—*When the sugar is inside, it'll be a little heavier, the sugar has a little weight.*—And when all of it has melted?—*It'll stay inside, the weight will stay the same, we just shan't be able to see it any more. It's no longer in a lump but like castor sugar. The weight stays the same.*—And the level of the water?—*It'll keep its place, the level will stay up. Oh, no, it'll drop back a little.*—Is there as much sugar as when it was in a lump?—*Yes.*—And when all of it has melted, will the weight remain the same?—*No, it'll be like half the lump; it gets thinner and thinner, some of it is always left but you can't see it.*—So will it keep its weight?—*Yes; no. It's a little heavier when it's in a lump.'*

LER (10; 9). The dissolved sugar *'will turn into powder.*—Shall we be able to see it?—*No.'* The sugar will make the water rise, but once it has dissolved *'the water will drop down again.*—Why?—*Because it's no longer heavy.*—How so?—*It's not as heavy as it was before. When it's in a piece it's heavier than in a powder.'* But when he is shown a glass with a lump of sugar beside it on one pan, and a second glass with the dissolved sugar on the second pan, he says: *'It's the same when the sugar has melted inside as if it's put by the side.*—And will the water stay at the same height?—*It's going to shrink a bit because there is less weight'.* Note the contradiction: the weight as measured on the balance remains constant, but the weight-force–which causes the water to rise – decreases because the lump is no longer in one piece! 'But you have just told me that they weigh the same on the balance?—*Yes, the water drops because there's less weight, but here* (on the balance) *it's the same weight on both sides.'*

VOI (11; 0) also starts out with the assertion that *'the water will drop back to the old level, because the sugar is not so heavy.*—And will it weigh as much as before?—*No, it'll be a little heavier because the sugar stays in the water.*—How?—*It turns into a very fine powder.*—Can we see it?—*No.'* The proof of the persistence of the sugar in the glass is that it can be recovered: *'We ought to dry it, and then we shall be left with the sugar that stays on the bottom.*—How will that look?—*It will be spread out in tiny little grains.'* Before comparing the weights on the balance Voi still thinks that *'the sugar has lost part of its weight',* but after the weighing he exclaims: *'Ah, you see, these tiny little grains inside do have some weight!'*

LIC (11; 7). Once the sugar has dissolved *'we shan't see it any more, but it will be in the water. It will be in the water like water ... when the sugar melts it changes into water.'* The glass with the dissolved sugar *'will weigh more* (than a glass of pure water) *because it has more water in it',* but *'that one with the sugar by its side will be heavier because the sugar will weigh more than the sugar water'.* Weighing: *'It's the same because this sugar* (the lump) *and that* (the dissolved sugar) *have the same weight.'* However, he predicts that the level will drop and even refuses to believe the evidence of his own eyes: *'I think it's dropped a little all the same, but you could hardly see it.'*

Now for some clear Stage III reactions, i.e. responses of subjects who accept the conservation of weight as a logical necessity but still deny the conservation of volume (Sub-stage IIIB):

SAC (8; 4) thinks that the dissolved sugar will turn into *'small grains'*, giving the water a sweet taste that *'will stay there for good.'* The water will rise but then *'it will drop back again because the sugar won't be there any more; it will all have melted.*—Will it disappear?—*No, it'll still be in the water, because when all the small grains are put together they make a lump.*—And will the water drop back all the same?—*Yes, because the lump will have gone.*—And what about the weight?—*It'll always be the same because we put one lump here* (by the side of the glass) *and another there* (in the water) *and the water in both glasses weighed the same.*—And will the water drop back again even though it's the same weight?—*Yes, when the lump is inside it will take up more space, and when it has melted it will take up less space.'*

After observing the actual level: *'That's because the sugar is inside, and that makes the water go up.*—But does it rise more when it's in a lump?—*No, because there are a lot of tiny little grains.'*

TOC (9; 1). The dissolved sugar has turned into *'powder'*. Toc thinks that the level will rise *'because the sugar melts and that makes some more water'*, but that *'it'll drop back again because all the sugar will have melted.'* However, the weight *'will remain the same, because even when it's all melted it stays the same.*—Shall we still be able to see it?—*No, once it's melted we can't see it any more because it's turned into powder.*—And with a lens?—*We shall see tiny little grains.*—And these two (one pan with a glass of sugared water, the other with pure water and a piece of sugar by its side)?—*They're the same, because you put a lump into this one and another one on that side, so it's the same.'*

After observing the level: *'It's because, though it's melted, the small grains have stayed in.'*

DID (10; 0). The dissolved sugar *'has turned into powder'*. The water will rise when a lump is dropped in *'because it takes up space'*, but it will fall back again *'because the sugar will have melted'*. The weight will remain the same *'because the sugar leaves its weight inside.*—How does it do that?—*It melts, but that leaves the weight as it was.'*

After seeing the level: *'It must take up room because the water hasn't dropped.'*

COM (12; 1). The dissolved sugar *'will turn into tiny little bits'*. The water will first rise but then it *'will drop back again, because when the sugar is whole, there's one place where the water cannot get in, but once it's melted it all mixes together and it'll drop back a little bit'*. The sugar water is heavier than pure water because *'there is the weight of the sugar on top of that of the water.*—Are you sure?—*I think so, because the weight can't possibly get out.'* If the water were evaporated *'all the sugar would stay behind.*—Could we make a whole lump of sugar out of it?—*Yes, by putting it all together into a lump, because it will get its hardness back that way.'*

All these responses pose two clear problems, namely how the child arrives at the idea of the conservation of the weight of the sugar, and why the same mental process does not lead him directly to the conservation of its volume.

It would seem first of all that simple identification is the most common explanation of conservation. Thus Ren said at the end of the test: 'It will be the same weight because the two glasses are the same, there is one lump to each'; while Toc predicted that the weight 'will remain the same', and Did explained that the sugar melts but leaves the weight 'as it was'. The best formulation of all was Lic's: 'It's the same (weight) because this sugar and that have the same weight.' However, in the case of the sugar as in that of the clay ball, these identifications are clearly the result of a rational construction, not the construction itself. The identity assumed by the child is, in fact, the equality of the whole to the sum of its parts regardless of any displacements. Now, nothing could be more complex than this final equality, as the whole history of grouping as we have followed it from Stage I to Stage IV serves to show: the identity (weight of the entire lump) = (weight of the scattered grains) is undoubtedly the product of a reversible composition, not *vice versa*.

The reader may recall that at Stage IIB (see Chapter 5, §2), the child constructs the substantial invariant by the simultaneous composition of displacements and divisions, but that he still refuses to apply that composition to the weight. In fact, the sum of the scattered particles in the glass of water will obviously seem to be equal to the initial lump if the child thinks of the substance alone, but the weight will not seem to remain constant while he still maintains that a separated particle is lighter than one combined with its neighbours into a single lump. The reason why subjects at Stage III arrive at the conservation of weight is, therefore, that they have adopted a system of compositions that ensures the equality of the total weight, the equality of the sum of its parts, and the equality of the individual parts, i.e. the conservation of the identity of any one part no matter where it has been shifted.

This is seen most clearly from the reactions of subjects at Substage IIIA who are still torn between the difficulty we have just recalled and the new grouping. Ren, for instance, still believed that the weight of the dissolved sugar would decrease 'because when the lump was together it was bigger, and the little grains are all over the water'. But then he went on to equalize the initial whole with the sum of the particles because 'the two glasses are the same; there's one sugar lump to each'. Similarly, Ler thought first of all that when the sugar is in a piece 'it's heavier than in a powder', but then changed his mind and said: 'It's the same when the sugar has melted inside as if it's put by the side.'

Similarly, Cha vacillated between the phenomenalist conception of weight, to which he returned in the end: 'It'll be like half the lump; it gets thinner and thinner, some of it is always left but you can't see it ... it's a little heavier when it's in a lump'; and equality: 'It's no longer in a lump but like castor sugar ... the weight stays the same.' Subjects at Sub-stage IIIB, by contrast, accept the conservation of the weight straight away and link it to the conservation of the substance precisely thanks to the double composition we have just mentioned. Thus, Sac defined the equality of the whole to the sum of its parts in a particularly lucid manner when he said that if 'all the small grains are put together they make a lump', whence he concluded that the weight 'will always be the same'. The words 'put together' and 'make a lump' are no longer mere descriptions of possible experiments but refer to real mental operations. Similarly, when Com said that the lump 'will turn into tiny little bits' and concluded that 'the weight can't possibly get out', he took for granted precisely what seems so doubtful to the younger children, i.e. that the weight of a 'bit' does not change with displacements.

But why is it that the equality or the identity of the weight of the separated parts should be grasped at this stage of development when it is so strongly resisted at the previous stage? It was not the formal mechanism of the composition that obstructed subjects at Stage II, since they applied this very composition to the substance. The reason why the equalization of the weights of the 'bits' is delayed is simply — as we saw time and again — that the act of 'weighing' is more closely bound up with egocentrism than the act of 'recovering', and that, as a result, the quality of the 'weight-force' remains more lastingly dependent upon the form and position of the ponderous object. Now, some of our subjects at Stage IIIA produce extremely striking examples of this opposition between weight conceived as a composable relation and weight conceived as the quality of an action. This was precisely why Ler, though accepting the conservation of weight when thinking of the reversible composition of the parts and the whole, nevertheless offered the same explanation of the rise in level as younger children are accustomed to do: he attributed it to the weight of the sugar lump. But his conception of weight was obviously quite different from theirs: his was a physical or quantitative concept divorced from the self and incorporated into an operational grouping, while theirs was still enmeshed in phenomenalist egocentrism. However Ler, too, contradicted himself when he explicitly attributed the alleged drop in level to a loss in weight, while admitting that, on the scale, 'it's the same weight on both sides'. At Sub-stage IIIB, by contrast, the intuitive idea of weight is finally discarded in favour of a quantitative conception: we may, therefore, conclude that the quantification of weight and hence the equalization of the differences or of the parts is due once more to the decentration of what used to be egocentric qualities and to their co-ordination into a grouping thanks to which they become detached from the self.

Hence it is only to be expected that this grouping or reversible composition should be reflected in further progress towards atomism. On the one hand, all subjects at Stage IIIA – except for Lic who thought that the sugar turns into water – adopted a form of atomism based on the schema of 'small grains', 'bits', 'powder', etc. On the other hand, since they endowed these elements with a constant weight as well as with a constant substance, their atomism tended towards greater operationality and acquired the character of a true schema of quantitative composition. Admittedly, this approach was ushered in during the previous stage, but it is only at Stage IIIB that it was adopted without reservation. This is why so many of our subjects derived their idea of the possible recovery of the original lump from a deductive atomism: if the water were evaporated 'all the sugar would stay behind', said Com, who also explained that it could be turned back into a lump 'by putting it all together . . . because it will get its hardness back that way'. Voi, for his part, explained that if the dissolved sugar were dried 'we shall be left with the sugar that stays on the bottom'.

But if the conservation of weight thus results from the reversible construction associated with atomistic composition, why does it not lead automatically to the conservation of volume? And if the total weight is thought to remain constant because it is composed of the sum of the weights of the particles that split up and became scattered during the dissolution, why is it that the total volume, too, is not considered equal to the sum of the partial volumes of the separate 'grains'? Now the very existence of this stage shows that the conservation of weight appears before that of volume, and that the discovery of the first does not lead directly to the discovery of the second. This is the more perplexing when we recall that, right at the beginning (Stage I), substance, weight and volume were relatively undifferentiated. In other words, it is only when the last two concepts become dissociated that they cease to be co-ordinated. Is the case of the dissolved sugar comparable in this respect to that of the clay ball?

We have been assuming that, in order to compose material transformations into coherent groupings, the child must reduce them to a system of reversible physical operations. These operations consist either of divisions (or recombinations) which correspond to operations with classes in logic or with cardinal numbers in arithmetic, or else of displacements which correspond to asymmetrical logical relations and to numerical ordinations, physical operations differing from the corresponding logico-arithmetical operations in that they replace logical succession and exteriority with temporal succession and spatial exteriority. It is thus that, in order to explain the conservation of the substance and weight of the melted sugar, the child thinks of the immersed lumps as dividing into increasingly smaller parts, and of the end products as being dispersed through the liquid, the group of displacements leaving the elementary objects invariant[1] and this group of divisions ensuring the constancy of their sum.

Are these operations as simple to perform with volumes as they are with weights? The arguments advanced by our subjects allow us to answer this question with some assurance. As far as the substance is concerned, the child starts to grasp at Stage II that no particles are lost and that one can always recover the original whole, much in the same way as he knows, from the age of one and a half to two years, that the sensori-motor object continues to exist even when it leaves his field of vision. But there is nothing to convince him as yet that the particle conserves its weight when it becomes displaced or scattered. At Stage III, he discovers that, if only he renounces the idea that weight depends on muscular effort, and assumes instead that it represents a relationship between objects, then the particles as such and also their sum conserve their weight. However, this does not yet mean that they invariably occupy the same space in the water; for as they become separated or scattered they may easily contract or expand. Thus Com contended that 'when the sugar is whole, there's one place where the water cannot get in, but once it is melted (= transformed into "tiny little bits") it all mixes together'; and Sac claimed that 'when the lump is inside it will take up more space, but when it has melted ("into tiny grains") it will take up less space'. Similarly, Ren said that the level 'will drop back again because the lump will have gone; so the water will take up the same space as before', because 'the little grains are all over the water'.

In short, the same arguments the child uses to deny the conservation of the volume of the clay ball during the displacement of its parts (deformation or division) will strike him as applying *a fortiori* to the volume of the sugar, because here the particles are not only reduced to the point of invisibility but are scattered throughout the liquid. Moreover, because sugar is permeable it seems to have no proper volume : its powder resembles a heap of sand that absorbs water without swelling, while the water itself is considered so elastic that its level is unaffected by the immersion of small

---

[1] For the relationship between 'object' and 'displacement group' see J. Piaget, *The Child's Construction of Reality*, Routledge & Kegan Paul, 1955, Chapters I and II.

corpuscles. Hence, before he can compose changes in the volume of solids and, *a fortiori* of liquids, the child must first learn to co-ordinate not only the dimensions of the object (see Chapter 3) but also its relative compactness, its *plena* and *vacua*, and to stop thinking in terms of compression or decompression. In short, the decentration of the egocentric relations and their grouping into an operational whole is much more difficult in the case of volume than it is in the case of weight.

§2. *The fourth stage (Sub-stages IVA and IVB):*
the conservation of volume, weight and substance

The conservation of volume is added to the other two invariants at about the age of eleven years. It is thus characteristic of a fourth stage comparable to the one we have examined in Chapter 3.

Let us begin with a few intermediate reactions (Sub-stage IVA), in which the conservation of weight is taken for granted from the outset, but the conservation of volume is still doubted, at least until the end of the test:

ERN (10; 0). The level *'will rise.—*Why?—*Because the sugar is heavier than the water.—*And when all of it has melted?—*It will drop back.—*Will there be some melted sugar in the water?—*Yes, everywhere, mixed with the water, all of it very tiny.—*And will the water drop back?—*Oh, no! the water will stay where it is.—*Why?—*Because the sugar is in the water, all very tiny.'* Weight: *'It'll weigh the same.'*

DUM (12; 0). *'The sugar has gone, it's vanished, it's all mixed with the water.—* Has it disappeared?—*The taste stays, the sugar is still inside.—*If we looked at it with a lens should we be able to see anything?—*We'd see small grains, but the lens would have to be very strong.'* The level rises after the immersion of the lump because *'the sugar takes up space, so the water is forced to rise.—* And when it's all melted?—*The water will drop back a little bit. It will not make as much volume in powder form as in a lump because it's not mixed with the water.'* Demonstrator drops the lump into the water: *'No, it won't come back to the same spot.—*Why not?—*Because the sugar is still inside.—*But you just said that it would drop back?—*I thought the* (dissolved) *sugar would take up less space. But it's not true.'* Complete dissolution: *'I was right.—*And what about the weight?—*The weight doesn't change.—*Why not?—*I've just seen that the water keeps to the same level even when the sugar has melted.'*

JAE (13; 9) says that the sugar *'weighs as much in the water as when it's dry'* but that, after dissolution, *'it'll drop back a bit.—*Why?—*First the sugar is dry, then the water gets into it and the sugar dissolves, so the water takes the place of the sugar.'* But then he adds: *'But I'm not sure, we'll have to see.'* Finally: *'The sugar takes up as much room as if it hadn't dissolved.'*

Next, responses by subjects who grant the total conservation of volume from the start (Sub-stage IVB):

FOE (9; 6). The water will rise because of the *'volume of the sugar.—And then?—It will melt.—*What does that mean?—*It'll turn into tiny little bits in the water.—*And will the water drop back?—*Oh no, it'll stay on top just like with the whole lump. The sugar has melted, but even so it takes up room in the water.—*And the weight?—*It'll be the same as the lump; it's a kind of liquid now, but there's as much liquid as there was sugar.'*

BURE (9; 9). The sugar will *'turn into crumbs.—*And the level?—*The water will stay on top, it still holds the same amount of sugar.—*Is the sugar still inside?—*Yes. The water will remain at the same height, because when you put the sugar in, it stays sugar. What I mean to say is that it's still the same thing even when we can't see it.—*And the weight?—*It's the same as well, because it's the same sugar all the time.'*

GER (9; 11). *'The sugar will melt, then it will split into two and next into tiny little scraps.—*Will it take up more space than a lump, or won't it?—*When it is in a lump it takes up a little more space than when it's in bits. No, no, that's wrong; it takes up the same space. I think the water ... Look, it's as if I took my cup and put a lump of sugar in, and I took another cup and put some sugar in it as well, but this time in grains, that would make the water rise just as much.—*And the weight when it's all melted?—*The same; it's as if we weighed a lump: it breaks up into bits, but when we put them all together again the small bits will make a big lump.'*

FEL (10; 11). *'The bits of sugar melt into tiny lumps we can't see but all the bits together, if they didn't fall down, that is, would make a big lump.—*What does the water do when we put the sugar in it?—*It goes up a little bit.—*And when it's all melted?—*It stays up; the sugar is in invisible parts in the water. It's all the same whether they are separate or in a piece; they occupy the same volume.—*And the weight?—*If we put some sugar into the water and weigh it quickly* (= before it dissolves) – *let's say we put in 200 grams. When the sugar has melted, it's still in the water so it has weight. If none of it evaporates it's still 200 grams: the sugar is still in the water.—*How can you tell?—*Because it always has the same taste.—*But some boys tell me that the taste stays but the weight decreases.—*If we weigh it – the proof is that if we take castor sugar and squeeze it into a lump, the two will weigh the same.'*

GIV (11; 2). The level will rise because *'the sugar takes up space even when it has dissolved because then it's in tiny bits.—*But you can't see any of them in the glass, can you?—*That's because the bits are so small.—*How much will it weigh once it's all dissolved?—*It'll always have the same weight: it's still the same bits inside but they're very tiny now.'*

ZUM (11; 6). *'The sugar will melt, the small grains will separate and we shan't be able to see them any more.—*And if we boil the water?—*We shall still have the sugar, it's like sea-salt.—*And the weight?—*It always has the same weight. The small grains spread out in the water, the sugar has stayed in the glass, and nothing has been taken out.—*And the level?—*When all the sugar will have dissolved the water will stay at the same height: the small grains take up the same volume as the large lumps.'*

ADI (12; 0). *'The sugar will melt.—*What does that mean?—*All the small parts are concentrated now, but when the sugar melts all the small bits separate, the water absorbs them. So we shan't be able to see them any more, but they*

106

stay in all the same.—How does the water melt the sugar?—*It makes the lump soft, it goes inside it. The grains aren't so concentrated any longer, they are pulled apart.*—Is it possible to have particles of sugar so fine that they won't melt any further?—*Yes, like the earth, it's made up of fine particles and they don't melt, but the sugar particles are much smaller, we can't see them.'* Weight: *'The weight will still be the same because the contents are still the same.'* Volume: *'It will still take up the same space.'*

SEL (12; 6). *'It'll rise when you put the sugar in, because it will take up more volume.*—And when it's all melted?—*It'll stay up. The small grains which make up the sugar stay in, but we shan't be able to see them any more; they've become transparent.'* He watches the sugar dissolve: *'Yes, I think it won't drop back. Those muddy bits* (the residue) *will stay at the bottom and if we mix them up they'll rise up again but they will still take up the same volume.*—And when it's all melted?—*It won't drop back.*—And where will the sugar be?—*Melted, dissolved. We shall only be able to see these rings* (the traces left by the sugar dissolving); *they make clouds.*—What are these rings?—*They're the sugar.'* Also: *'the quantity remains the same.*—And the weight?—*It won't change either. It's as if we crushed a stone into dust. It's the same thing, the same weight because the stone is made up of small grains of sand.'*

DRE (12; 9) thinks that the level will stay up: *'It takes up the same volume because the sugar is inside all the time.*—How so?—*It stays in the water but in a large number of small bits. It takes up the same room, it has the same volume as the cube.'*

The progress of these subjects does not merely lie in the discovery of the conservation of the volume of the melted sugar, as the last of the three invariants, but also and above all in the method of composition by which they arrive at the new principle of conservation and in the special type of atomism they introduce. Let us look at the last point first.

The pre-atomism of Stage I is only a perceptive representation of the visible 'crumbs' at the point of dissolution, together with the belief that they cease to exist once they can no longer be observed. At Stage II this primitive atomism is extended by the notion that, after dissolution, the crumbs or grains continue as invisible particles which account for the persistence of the taste. However, since these atoms are thought to be devoid of either weight or volume they merely reflect the intuitive belief that objects can always be retrieved. This belief entails an implicit quantification of the substance. Stage III produces a notable advance: the 'grains' are subjected to a second quantifying composition thanks to which each one of them is endowed with weight, and the sum of their weights is equated to that of the original lump. But this mode of composition does not yet extend to the volume and simply constitutes a method of simple addition or combination of the parts. Now the real achievement of subjects at Stage IV is that they succeed not only in generalizing this schema by applying it to the volume of the elementary grains, but also in incorporating it in the wider schema of compression and decompression, which accounts for changes in the contours of the sugar as it is transformed from a cube into a kind of transparent syrup, into 'rings', 'clouds', etc. circulating freely through the water.

In fact, children at Stage IV no longer treat such substances as sugar, stones, clay, etc. as simple conglomerates of fused (= solids) or separated grains (= powder, dust, etc.), but as differing in compactness, hardness, resistance or density thanks to the underlying schema of compression ('tight') or decompression ('loose'). They accordingly distinguish between (1) what we shall call the global volume, i.e. the volume of the combined grains plus that of the interstices, and (2) what we shall call the total or rather the total corpuscular volume, i.e. the sum of the volumes of the individual grains but not of the interstices. Thus when confronted with the expansion of the maize seeds, which we shall be considering in the next chapter, they realize full well that though the global volume increases, the elementary particles of the flour do not increase in volume but simply draw apart under the influence of the heat. Now, in the case of the sugar it is precisely the confusion between the total corpuscular volume and the global volume which explains the difficulties and the non-conservation characteristic of the earlier stages. Thus when Jae (Sub-stage IVA) still believed that the level of the water would drop because 'first the sugar is dry, and then the water gets into it and the sugar dissolves, so the water takes the place of the sugar', or when Com (Stage IIIB) said, 'When the sugar is whole there's one place where the water cannot get in, but once it's melted it all mixes together and it'll drop back a little bit', they obviously failed to equate the sum of the volumes of the dispersed grains with the total volume, and this precisely because they failed to distinguish the total from the global volume. Moreover, the reason why they could not distinguish between these two notions was plainly their failure to realize that though the global volume of the grains varies with their compression or decompression, the total corpuscular volume remains constant. By contrast, when Adi (Stage IVB) explained that in the initial lump 'all the small parts are concentrated now, but when the sugar melts, all the small bits separate ... The water makes the lump soft, it goes inside it. The grains aren't so concentrated any longer, they are pulled apart', the schema of compression and decompression enabled him to distinguish the variable global volume from the constant total volume, both of the sugar as a whole and also of each particle. Similarly, when Fel said 'the sugar is in invisible parts in the water.

It's all the same whether they are separate or in a piece; they occupy the same volume'; or when Zum said 'the small grains spread out in the water ... the small grains take up the same volume as the large lumps'; or when Sel explained that the dissolved lump turns into 'rings' or 'clouds' and that 'it's as if we crushed a stone into dust ... because the stone is made up of small grains of sand'; they were all applying the schema of compression and decompression and making a clear distinction between the total corpuscular volume and the global volume: the latter varies when the parts separate, expand, form clouds, etc., while the former remains constant. As for those subjects who made no explicit reference to the schema of compression and decompression, it often suffices to ask them why the sugar melts when a pebble does not,[1] for them to come up with the answer that the pebble is made up of 'crystals of sand which become stuck together after a dry spell' (Rog, 10; 1), or that the stone is 'more compact', 'fuller', 'more closely packed' and 'harder' (Ber, 12; 0, and many others).

In other words, the total corpuscular volume appears in the wake of a correlation of the global geometrical contours with the relative fullness of their contents, based on the schema of compression and decompression: it is seen to be constant because it is always equal to the sum of the constant volumes of all the particles. The global volume, by contrast, changes upon dissolution, only the volume of each individual grain remaining invariant.

To put it more concisely, the global volume is altered by compression or decompression, but these alterations do not affect the volumes of the fragments or the total volume resulting from their combination. Now, this interpretation is open to one serious objection. We have been assuming, in the case of the clay ball no less than in that of the sugar, that before they reach the stage at which they grasp the conservation of volume, our subjects assume that substances have a kind of passive elasticity such that every deformation produces an expansion or contraction of both the particles and the whole, and we have tried to show that the grasp of the conservation of the total physical volume depends on the realization that no one particle changes volume when it is separated from the whole, and that no one 'lump' of matter changes concentration when it is deformed or divided. It may therefore be argued that the belief in changes in concentration is nothing but the schema of compression and decompression, the more so as this schema appears at precisely the same time as the conservation of solids. However, it is very easy to meet this objection, which is a purely semantic one. In fact, the assumed instability or pseudo-elasticity

[1] See Chapter 8, §4.

of matter is merely a consequence of the belief in an intuitive process at the end of which neither the grains nor the lumps have totally or globally constant volumes: that belief is therefore both a negation of permanent concentration and also of the invariance of the volumes of the elements, i.e. it rests on a confusion of the total corpuscular volume of the parts with the global volume. Now, without the requisite invariants, the schema of compression and decompression cannot be applied, and there can be no composition. With that schema, by contrast, the child has an operational method of accounting for the convergence or separation of elements with a constant global and total volume. In the case of such solids as the ball of clay, this leads to the idea of a permanent concentration and hence of the invariance of volume, both global and total. In the case of the dissolved sugar on the other hand, it allows him to consider the volume of the particles as being invariant, i.e. to equate the total volume to the sum of the partial volumes, and to treat the global volume as varying with the convergence or separation of the parts. In brief, the characteristic notions of these two levels are as far apart as undifferentiated intuition is from the operational schema, i.e. from the method of reversible composition.

Once the operational schema has been adopted, the conservation of volume becomes self-evident and, moreover, inseparable from the atomistic compositions we have just described. It is obvious, first of all, that such simple identifications as Bure's 'When you put the sugar in, it stays sugar. What I mean to say is that it's still the same thing even when we can't see it', are the end results of a complex operational argument involving the operations of division (equality of the whole and its parts) and displacement (preservation of the total corpuscular volume during decompression), which other subjects invoke explicitly (Giv, Zum, et al.). In particular, some of them stress the reversibility of the operations they employ by explicit reference to the inverse operations. Thus Ger, after explaining that the original lump splits into two and then into small particles, and having postulated the equality of the whole to its parts ('It's as if I took my cup and put a lump of sugar in, and I took another cup and put some sugar in it as well, but this time in grains, that would make the water rise just as much'), went on to describe the inverse operation: if all the fragments of the original lump were put together 'the small bits will make a big lump'. Similarly, Fel: 'The bits of sugar melt into tiny lumps we can't see, but all the bits together, if they didn't fall down, that is, would make a big lump' (cf. his arguments about the weight). In short, because both are based on the three operations of division, displacement and compression, and their inverse, the atomism characteristic of this level and the conservation of volume constitute one and the same explanatory system.

But there is yet another way of justifying conservation, namely by the extrinsic co-ordination of the three invariants (substance, weight and volume). In fact, it is enough to demonstrate the conservation of any one of the three invariants by means of an atomistic composition or by one of the methods of identification we have described, for children at Stage IV to extend this proof to the other two directly thanks to the logical nexus they have established between them. Thus Dum justified the conservation of weight by that of volume: 'The weight doesn't change,' he said; 'I've just seen that the water keeps to the same level even when the sugar has melted.' Others use the conservation of substance in its most qualitative form, i.e. taste, to deduce the invariance of the weight and volume: 'Let's say we put in 200 grams,' said Fel. 'When the sugar has melted, it's still in the water, so it has weight. It's still 200 grams.' And when he was asked for proof, he said: 'Because it always has the same taste.' Others, again, justify the conservation of volume by the conservation of substance or weight, etc.

Thus, in the case of the sugar no less than in that of the clay ball, the concepts of substance, weight and volume, which were originally confused in the same egocentric and phenomenalist schemata and which, once dissociated, developed in separate directions, are finally co-ordinated into a coherent overall system. Nothing is more characteristic of this process than the co-ordination of weight with volume. We saw that during the preceding stages this relationship was still far from being reciprocal: when the child discovered that the level of the water did not drop back after dissolution he often concluded that the weight had remained constant, even if he had not realized this fact before; but when he saw on the balance that the weight had been conserved he never concluded as to the permanence of the volume. In other words, not only does the conservation of weight appear spontaneously before that of volume, but, even during the experimental demonstrations at the end of the test, the second of these invariants often leads to the first but not *vice versa* (the apparent exceptions are due to residual phenomenalism: the child simply believes that the level will not drop once it has risen because there is nothing to pull it down). At Stage IV, by contrast, any one of the three invariants is seen to imply the others, because the grouping of all the operations involved has been completed.

§3. *Conclusions*

Three parallel series of facts thus appear during the four stages which were examined in Chapters 4–6.

There are first of all the notions on which the child bases his predictions: absence of all conservation, followed by the conservation of the substance, the conservation of weight, and finally the conservation of volume, every one of these three invariants becoming integrated with preceding ones until there is the total conservation characteristic of the final phase of this development.

Second, there are the explicative notions which enable the child to picture matter by atomistic schemata based on conservation or non-conservation. Non-conservation is bound up with a representation taken from the child's direct experience that the sugar lump splits into 'crumbs' and eventually disappears in the water. The conservation of matter goes hand in hand with notions ranging from pure liquefaction of the melted sugar to its persistence in the form of invisible and imponderable 'grains'. The conservation of weight brings an advance in atomistic composition inasmuch as the equality of the initial whole to the sum of its parts is no longer applied to the substance (taste) but also to the quantification of the weight. As a last step, the conservation of volume brings a further advance in correlation, thanks to the schema of compression and decompression and the distinction between global volume and total corpuscular volume.

Third, the reactions produced by the experimental demonstrations at the end of the test (of the constancy of the weight and level) can be fitted into a similar series. Failure to grasp conservation and atomism is reflected in complete failure to appreciate the import of these experiments; the child simply fits the data into his egocentric *cum* phenomenalistic approach, and hence denies the conservation of matter. Once he discovers the latter and begins to move in the direction of atomism, i.e. begins to think in terms of substantial but imponderable grains without volume, he also pays greater heed to the experiments but fails to reach complete co-ordination: from the constancy of the level he deduces the conservation of the volume and hence of the weight, but not *vice versa*. Stage III sees two advances in the relationship between reason and experience: on the one hand the child anticipates experience by predicting the constancy of the weight, and on the other hand he bows to the experimental demonstration of the conservation of the volume and co-ordinates it with that of the weight, but by induction and not by anticipative deduction. Finally, at Stage IV there is a complete reversal of what happens at Stage I: the child deduces and co-ordinates the three forms of conservation, and all the experiments do is confirm his *a priori* conclusions.

What are we to make of the relationship between these three

series of facts? Since atomism and the construction of the three invariants result either from an experiment, or from an *a priori* deduction, or again from a combination of the two, we must, in order to grasp the general mechanism of conservation, begin with an analysis of the relationship between experience and reason. Now this relationship seems particularly paradoxical and incoherent in the third series, and this precisely because the construction of the three invariants is not the result of pure experience or of pure reason, but of a complex combination of both. That the principles of conservation and of atomism do not have a pure *a priori* origin is quite plain from their evolution. Thus all our subjects know that the dissolved sugar leaves a more or less lasting taste in the water, and this fact alone suggests to them, as soon as they have gone beyond egocentric phenomenalism, the idea that some of the substance must be conserved. Moreover, it is quite possible that older children have observed that the level of, say, a cup of coffee remains constant after it has been sugared. Finally, though the dissolved sugar is transparent and its molecules invisible, it nevertheless remains a fact that once he has adopted atomism the child extends it directly to the vanishing sugar grains. For all that, it is clear that our experiment would not have led him to the idea of complete conservation or to atomistic compositions had deductive factors not helped to structure and complete the perceptible data. This is borne out by the fact that, as each new invariant is constructed (Stages IIB, IIIB and IVB), the child treats it as an *a priori* necessity, in complete contrast with his earlier, purely phenomenalistic approach. Now, the appreciation of logical necessity in the child, as in the scientist, goes beyond the experimental data and bears witness to the intervention of deduction. Indeed, how can experiment alone explain the quantification that appears with each new conservation? It is obvious that the child has never tried to ascertain the total conservation of the substance, weight and volume of the sugar by careful weighing and measuring experiments of his own. Hence his new conviction is more than the result of pure observation: it is the product of deductive certainty. Now, quantification is not a characteristic that is simply added to the principles of conservation once the latter have been constructed, rather is it the very crux of the method of composition leading to that conservation. Finally, to say that such experimental data as persistent taste, constant levels, or the conversion of the lump into 'grains' suggest the invariance of matter, weight and volume, is merely to say that these data may serve as elements of a schematic elaboration, but it is quite obvious that they can only do so provided a deductive construction helps to structure and complete them by fitting them into a system of coherent operations. This is proved by the fact that the same experiments make no impression on children at Stage I, who are still incapable of deduction, and also by the fact that the necessary co-ordinations appear only gradually during the following stages.

Conservation, like atomism, thus demands a co-ordination of experience with deduction. Let us follow both step by step through our four stages, remembering that they are inseparable in practice. During Stage I, the child engages in an egocentric assimilation of the external data to such endogenous schemata as the taste-quality, the weight-force, subjective growth, intuitive voluminosity, etc.; as for experience, it only manifests itself in its immediate and perceptive form, i.e. as phenomenalism. The fusion of egocentrism with phenomenalism is thus the most primitive link between the most superficial awareness and the most superficial experience. At Stage II, the subject's own activity begins to be differentiated from external experience and the two can then be co-ordinated instead of impeding each other: taste becomes a lasting quality and suggests the existence of permanent substances, much as the final weighing and measuring experiments suggest the conservation of weight and volume. During Stages III and IV, the differentiation and co-ordination of deduction and experience becomes accentuated: the first permits the quantification of weight followed by that of volume, while the second confirms the child's anticipations. In short, it would seem that as soon as deduction has been applied in one sphere (the substantial invariant) experimental induction can be applied to the others (weight and volume) until, finally, the entire system becomes deductive and hence closed.

In other words, the correct reading of experience and reason must be based on a single system of relations involving both induction and deduction, the inductive composition being constructed step by step, and the deductive operation combining the relations into a complete whole. While the relations are not yet co-ordinated, as happens at Stage I where every attempt at composition ends in unsurmountable contradictions, neither deduction nor a correct reading of the experimental data is possible, but as the mental operations become differentiated and the data are no longer distorted by the experiencing self, correct deduction and induction follow as a matter of course. It is, therefore, an oversimplification to say that experience supplies the relations while deduction co-ordinates them: the subject's own actions intervene in the elaboration of the relations, helping to replace irreversible processes with reversible operations, and the co-ordination of the relations, though not independent of their contents, becomes confined to the composition of the experimental transformations into an effective grouping. What must be stressed is that the development of the links between deduction and experience proceeds from a type of phenomenalism that is unconsciously centred on the self to the construction of a group of physical operations, the initial situation reflecting a complete failure to apply objective co-ordinations because the reference system is still based on egocentrism, and the final stage being characterized by a logico-mathematical *cum* experimental co-ordination, because the initial system has become decentred to make way for a general composition fitting the activities of the self into a universe of reversible transformation. The relations between substance, weight, and volume established by children at Stage I cannot, in fact, be composed into a single deductive system because all three have remained simple qualities both subjective and bound up with direct perception : hence they can only give rise to postductive juxtapositions or to transductive fusions. By contrast, deduction, and with it induction or the possibility of accounting for experience, begins with a construction that lends itself to both associative composition and to reversibility, i.e. a composition that tends to assume the form of groups.

What precisely are the groups our subjects employ to establish the conservation of an atomistic structure of dissolved substances? As we saw time and again, they involve the replacement of intuitive relations with operations (divisions and displacements). Hence the resulting atomism, whose gradual development we have followed, is never a collection of representative schemata : it is a system of compositions, and while the imagination provides these compositions with a symbolic substrate, it also bows to their demands, so much so that operationality gradually gains the upper hand over representation.

Now the form of the group or grouping resulting from the gradual structuring of deduction is of great importance to the general theory of the relation between reason and experience. We saw that the content of the groupings we have been discussing is invariably derived from the source of experience : the relations between volume, weight and quantity of matter, the idea of atomic grains, the transformation by division, the displacements, compressions, etc. are all imposed by the facts as such. What, then, is the share of the mind in the elaboration of physical operations? As the child advances from phenomenalist egocentrism to grouping, i.e. from the centration on his own actions to the co-ordination of all the possible views, he does not start to ignore action : an operation, i.e. the element of any one grouping, is always an action but one that has been decentred from the self and that, having become reversible, can be composed with any other. Hence, no matter how inductive the progress of experience, and no matter how empirical the composition of the experimental data, the child's action will always be based on that logical necessity which deduction alone can provide and which is reflected in the composition of operations with other operations or with their inverse within systems that are both closed and open to all combinations. The unity of action and object is therefore as close at the end of the day as it was at the beginning: but instead of reducing the universe to himself the subject ends up as part of the universe, which he can at last co-ordinate, thanks precisely to the incorporation of the external transformations into a system of reversible operations.

PART III

Compression, Decompression and Density

The Expansion of a Maize Seed and of a Column of Mercury[1]

In the last section, we remarked upon the emergence of a schema of composition different from that of division or simple displacement, namely the schema of compression and decompression. We must now look at it more closely and, to ensure continuity between the present study and the preceding analyses, we shall begin with changes in volume that leave the weight and quantity of matter unchanged. Two experiments seemed particularly suitable, the first of which proved preferable to the second. In it, a maize seed is placed on a heated asbestos plate and the child is asked to watch it 'pop'. He is then asked whether the weight and the quantity of matter have remained unchanged and why the volume has increased. The disadvantage of this method is that the conservation of weight and of matter is not complete because of the slight evaporation of the water in the seed, though it is simple to tell whether the child denies the invariance for pre-logical reasons, or whether he appreciates that the decrease of matter is due to evaporation. As for his explanations of the changes in the global volume, i.e. of the expansion or decompression, they may, as with the sugar, be based on an atomistic composition, which again raises the question of whether such atomism is characteristic of a particular stage or whether it occurs at all stages.

In the second experiment, alcohol or mercury is allowed to expand in a thermometer tube and the child is again asked if the weight or quantity of matter has changed during the expansion and also to explain the expansion. The disadvantage is that the child may be unfamiliar with the structure of the thermometer and hence may focus his attention on secondary problems, and also that the closed system constituted by the tube fosters the idea of conservation artificially. Hence we have merely used the thermometer test as a control experiment, and shall mention it only in passing.

The expansion of the maize seed produced reactions that could

[1] In collaboration with Nelly Gruner.

all be fitted into the four stages we have described. During Stage I both matter and weight are thought to increase with the expansion, which is thus treated as a process of immediate and absolute growth. During Stage II there is conservation of matter but not of weight which is generally thought to decrease as a result of the expansion or swelling of the seed. At Stages III and IV the discovery of the conservation of weight plus the discovery of the increase in volume lead an ever-greater number of subjects to an atomistic explanation : at Stage III the swelling is attributed to the particles themselves, each of which is believed to expand separately; Stage IV introduces the idea of the decompression of elementary seeds of constant volume. The thermometer experiment confirms the succession of these stages, but with it the child can do no more than hint at his grasp of atomistic composition by speaking of 'drops' or 'balls'.

§1. *The expansion of the maize seed: the first two stages*

Stage I and Stage II responses must be analysed not only to check the conclusions reached in Chapters 4 and 5, but also because they cast a great deal of light on the compression schema : while there is still non-conservation of substance or weight there can be no clear idea, and above all no atomistic explanation, of the expansion. The idea of elementary grains can, of course, exist even at Stage I, but only as a purely perceptive notion. Moreover, at this stage, the child is still inclined to think that the number of these grains increases with the expansion. At Stage II, by contrast, the substance and possibly the number of grains as well are conserved, but the weight is thought to vary.

Here, first of all, are several Stage I responses :

NOS (8; 0). 'Will the seed keep its size?—*No, it's going to get bigger, it'll swell.*—Why?—*It's the heat. It'll burn a bit, and then it'll swell up, just like when you burn your finger.*—How do you explain that?—*There are grains inside, the heat makes them come out.'*—The seed bursts: 'Is it still the same weight?—*It's got heavier.*—Why?—*Because it's bigger.*—If you look at this one (an unheated seed = A) under a microscope what do you think you will see?—*Small grains.*—And this one (A' = the expanded seed)?—*This one has more grains.*—How can you tell?—*Oh, no; they have the same number but they have grown bigger.*—Is there more flour in this one (A') than in that one (A) or are they the same?—*There is more.'*

WEN (9; 0). 'What is going to happen next?—*The seed is going to swell up like rice.'* The seed bursts: 'Why?—*It swells up because the fire makes it bigger.*—Will it weigh the same as before?—*No, this one (A') weighs more: it is heavier.*—Why?—*Because it's grown bigger, it's got more stuff than before.*—More of what stuff?—*Flour.*—Did we add any?—*No, but the fire did.*—How?—* . . .*—What happened to the seed?—*The things inside it* *have grown bigger.*—What things?—*Grains of flour, the grains from the seed. Before they were all tiny, now they are big, so they must weigh something.*—But are there as many grains inside as before or not?—*There are more than before.*—And is the big seed heavier than before or not?—*It's heavier because it's bigger. When I was small I weighed less.'*

SOR (9; 6) after the seed has burst: *'It's blown up inside.*—Why?—*The heat rose up and made the skin burst under the pressure.*—Under the pressure of what?—*Of what was inside.*—What was inside?—*Corn.*—What do you think has happened to the weight?—*It got heavier because it grew with the heat. It pushed out. It got bigger.*—How?—*Because what was inside wanted to get out. It's like a seed, it germinates more quickly when it's hot.*—Is there as much of it as before?—*There's more flour.*—And where does the extra flour come from?—*It has added itself and grown bigger. The heat has pushed it out.'*

We see at once that none of these subjects has the least idea of the conservation of weight: they all think it quite natural that the seed should gain weight with increases in volume, because they quite generally believe that the two are proportional, i.e. that the substance itself increases upon expansion. Now, no matter whether that substance is said to be continuous (Sor's 'flour') or discontinuous (Wen's 'grains') its quantity is always thought to increase: 'There's more flour,' said Sor, because 'it has added itself and grown bigger', by which he did not merely mean to say that it had expanded but also that it had 'pushed out' like a growing plant. Similarly those subjects who mentioned a granular structure assumed that the number of 'grains' increased with the expansion: 'This one has more grains,' said Nos; 'there are more than before,' said Wen. Or else they contend that the elementary grains themselves have increased in substance: 'they have grown bigger,' Nos also said, not realizing that he had merely shifted the problem. In other words, he refused to choose between two hypotheses that struck him as alternatives introduced artificially by the investigator.

Quite generally, all these subjects consider it self-evident that, upon expansion, the seed should increase in substance and weight, much as a plant or animal gains weight and stature when it grows. That is what Wen means when he compared the maize seed to his own body: 'When I was small I weighed less'; or what Sor meant when he said: 'It grew with the heat. It pushed out.' Nos even went so far as to compare the expansion of the seed to a blister: 'It'll swell up, just like when you burn your finger.'

In brief, the expansion is mistaken for an increase in substance, because children at Stage I simply translate the phenomenal impressions of their direct perceptions into the language of egocentrism. Children at Stage II, by contrast, begin to dissociate and compose the experimental data: while the weight is still believed to vary, the quantity of matter begins to be conserved much as it was in the case of the clay ball and the sugar. But, interestingly enough, the weight is now generally thought to decrease.

MEY (7; 6). *'Oh, the seed's grown bigger.—*And its weight?—*It'll weigh less now that it's opened up.—*Why?—*Because there's more air inside.—*How so?—*It's the air from the fire, the smoke gets into the grain.—*But is it the same inside as before, or not?—*It's the same, but it's lighter.'*

SIER (9; 6). *'Inside the grains there is some very fine powder.'* The seed bursts: 'What's happened?—*It's grown bigger.—*And what shall we see on the balance?—*That it's become lighter.—*Why?—*Over here* (A) *it's all tight, so it'll weigh more, but over there* (A') *it's completely opened up. It's burst, so it'll weigh a little less.—*Is there the same amount of flour inside?—*A little less; no, it's the same amount.'*

GRIL (9; 9). 'What's going to happen?—*It's going to get bigger.'* The seed bursts: *'It's burst.—*Why?—*It's like a flower, like a bud that bursts open.—*And its weight?—*It will get lighter.—*Why?—*I don't know why, but you can see for yourself.—*Is there more or less inside?—*It's the same.—*Of what?—*Small grains; over here* (A') *the small grains have been changed into dough.—*Why does it get lighter when it's bigger?—*It's been blown up; the heat always does that.'*

MAT (10; 0). *'The heat has blown them up. These seeds breathe the heat in.—*How so?—*The heat gets inside them and takes up room.—*But how?—*The grain has pores through which the heat can pass.—*Has it stayed the same weight?—*This one* (A) *is heavier, and that one* (A') *is lighter.—*Why?—*The air weighs less than the seed; it takes away part of the weight.—*Why?—*That one* (A) *has not been opened up by the heat; when it opens, it loses part of its force.—*But how is it that this one (A') is bigger and lighter?—*The heat takes up space, the stuff inside has been blown up.—*What stuff?—*Skins packed tight against one another.—*And what does the heat do?—*It pulls them apart.—*And if we could count them, would there be as many in here* (A) *as in there* (A') *?—*Yes, it's the same number because it's the same grains. One has been blown up and the other is closed, but it's the same.'*

We see that all these children, unlike those at Stage I, postulate the conservation of substance: 'It's the same, but it's lighter' (Mey); 'it's the same amount' (Sier); 'it's the same' (Gril); and above all, 'it's the same number because it's the same grains' (Mat). However, they all refuse to grant the conservation of weight: to most of them the expanded seed becomes 'lighter', which does not mean that they think its relative weight decreases while its absolute weight remains constant, but rather that its absolute weight changes. More precisely, these subjects have not yet grasped the distinction between weight and density. By what system of reasoning, then, do they arrive at the conservation of the substance while refusing to grant the conservation of weight? The first part of the question is easily answered, but the second is not because it raises the entire problem of compression and density.

In fact, the child need merely apply methods 1 and 2 (see the end of Chapter 1) to discover that the substance of the seed is invariant. Thus once he assumes that the seed is made up of flour, dough, 'small grains', or 'skins', etc., he realizes that the substance remains constant provided every 'skin' or small grain retains its identity during displacements (method 1) and that the number of grains remains constant (method 2). When Gril said that the exploded seed contains 'the same' substance as the unheated seed because the 'small grains have been changed into dough' he was therefore applying method 1; and when Mat said 'it's the same number' he was explicitly using method 2.

However, is the fact that he grasps the conservation of substance when the volume expands so dramatically not a sign that the child has also acquired an intuitive understanding of density, i.e. of the relationship between mass and volume? Moreover, when he thinks that the weight decreases with increases in volume, does that not mean that he is thinking of the relative rather than the absolute weight, and that our own interpretation is quite false? In other words, is he not applying the schema of compression and decompression—as witness his use of such terms as 'tight' or 'blown out'—to reconcile the conservation of matter with the increase in volume? In our view, this interpretation is quite invalid: the terms the child uses, far from being signs of a real composition (except for the conservation of substance), merely show that he has incorporated his discovery of the expansion of the seed into his intuitive or pre-operational schema based on the instability of matter.

In fact, what does the child really think when he argues that the burst seed has become 'lighter'? Mey, for example, asserted that the seed 'will weigh less now that it's opened up', and this 'because there's more air inside'. He in no way suggested that what had changed was the relative weight, for in that case he would simply have said something like: 'It's the same weight but it's bigger, so it's become lighter.' Similarly Sier was convinced that the scales would reveal an absolute decrease in weight: 'Over here (A) it's all tight, so it'll weigh more, but over there (A') it's completely opened up. It's burst, so it will weigh a little less.' Gril, too, did not consider the alleged decrease in weight a decrease in density but treated it as an absolute loss, and one, moreover, that struck him as being quite obvious: 'I don't know why, but you can see for yourself.' Finally, Mat came out with a fuller explanation, when he stated explicitly: 'The seeds breathe the heat in', and 'the air weighs less than the seed; it takes away part of the weight', because when the seed opens 'it loses part of its force'. There could be no clearer way of demonstrating that, for children at this stage, weight is not yet subject to the rules of composition defining the whole by the sum of its parts, but that it remains a force and can therefore be evaluated in a purely intuitive way by the pressure it exerts on the hand. Considered in this way, a compact seed does indeed seem heavier, i.e. more forceful, than a seed filled with air.

In other words, the use of such expressions as 'tight' and 'blown up' to describe the seed before and after heating in no way attests to the use of a rational schema of compression and decompression, i.e. of one by which the weight and total corpuscular volume of the whole remain constant and only the global volume changes. In fact, it is not until Stage III that the child grasps the constancy of the weight, and not until Stage IV that he comes to distinguish the total corpuscular volume from the global volume. At the present Stage we cannot therefore speak of an operational schema of compression and decompression, but simply of an empirical 'swelling' schema, according to which the substance alone is constant but becomes filled with air and hence changes its weight and volume. Thus Mey declared that the seed had 'opened up' because 'there's more air inside', and that the air comes from the fire: 'the smoke gets into the grain.' Sier, for his part, simply stated that the seed had 'burst' and that 'it's completely opened up'. Gril explained that the 'small grains have been changed into dough', because 'it's been blown up; the heat always does that'. Mat, who thought of the seed as an onion formed of a succession of layers believed that the heat pulls apart 'skins packed tight against one another'. Now, none of these responses was part of an atomistic schema based on divisions and displacements and in which compression and decompression are the direct results of the spatial arrangement of the particles. Instead, the 'swelling' mentioned by these children reflects a belief that the seed has a pseudo-elastic or unstable texture, i.e. it is still founded on an intuitive and perceptive schema, as witness the fact that the 'swelling' is thought to alter the weight of the seed.

§2. *The third stage: conservation of weight without conservation of volume*

We saw that at Stage II the child believes that the substance alone is conserved, while the weight decreases and the volume increases. The resulting dissociation of the weight from the substance is a habitual one, for it also occurs at Stage II in the clay ball and sugar tests. By contrast, the dissociation of the weight from the volume is a new feature, and one that we shall meet again in the cork and pebble experiment (Chapter 8). It becomes even more marked during

Stage III, when the weight becomes constant and hence allows of the rational composition of the ponderous particles, while the global volume is thought to increase and, with it, the volume of every particle (as already happens at Stage II). We therefore have progress in the sphere of weight relationships, but not in that of volumes, with the result that the schema by which weight and volume are correlated does not yet go beyond the intuitive idea of a simple 'swelling'.

Let us begin with two intermediate cases, still wavering between the non-conservation and the conservation of weight (Stage IIIA):

BERT (10; 0). 'Why has this seed grown bigger?—*It's the heat, it heats the seed and then it opens up.*—But how does it get bigger?—...—And what about its weight?—*It'll stay the same, or maybe it'll get heavier. When it's small it doesn't weigh very much, but when it's big it may weigh more.*—Is there more maize inside?—*No, it has as much maize as before.*—So why should it get heavier?—*Because it's bigger. Oh, no, it's the same weight, because over here* (A) *it's inside and so it's smaller, and over there* (A') *it's opened up but it's still the same weight.'* The reason why the seed 'swells up' is that *'there are very small grains in this one* (A).—And what about that one (A') ?—*It also has small grains.*—The same number?—*No, more, or perhaps it is the same.*— Which is right?—*It's the same. It's the same number, but over here* (A) *it's inside, it's smaller, and over there* (A') *it's opened up, the grains have become bigger.'*

CLAV (11; 0). A' *'is lighter.*—Why?—*It draws apart, it blows up like a balloon and then it cools down and hardens.*—How?—*When you heat it, it becomes softer, it draws apart, and so the gas* (hot air) *blows it up.*—Is it the same weight?—*It's become bigger, and so it's lighter.*—Why?—*Over here* (A) *it was all tight and there* (A') *it's grown bigger.*—And is there still as much maize or not?—*It ought to be the same, it's simply grown bigger.*—How?— *It's been blown up.*—And so it's lighter?—*Oh, no. It's the same. Over here* (A) *it's all packed together and over there* (A') *it's grown bigger, it's been pulled out. The parts have remained the same but they have grown bigger and so they take up more space. Over there* (A) *they're all squashed together, and here* (A') *they're side by side but pulled apart. They're the same weight but bigger.'*

Now for some unequivocal Stage III responses (Sub-stage IIIB):

ROU (10; 0). 'What is this maize seed made up of?—*It's got flour inside.'* The seed bursts: *'It's bigger.*—Is there more flour inside?—*No, it's the same, because it's still the same seed.*—Is it heavier than before or lighter?—*It's the same.*—Why has it grown bigger?—*The flour has burst out.*—What is the flour made of?—*Of small grains.*—If we could count them, would there be as many when it's burst as there were before?—*Yes.*—So what happens when the flour bursts?—*The small grains swell up.*—How?— *The heat makes them fly apart, they get blown up.*—What does that mean?— *That they get bigger.'*

ALT (10; 0). *'The seed has grown bigger.*—And its weight?—*It's the same weight, it's just burst.*—But why is it the same weight?—*None of the bits have dropped off.*—What is the seed made of?—*Of flour or powder.*—Is there more flour now?—*It's the same; it's bigger, but it's the same as before.*—Why is it bigger?—*When it bursts it swells up, it gets bigger and thinner.*—What becomes bigger and thinner?—*The flour. The heat blows it out, its skin bursts.*—Is this (metal) bar made up of powder?—*No, powder is made up of small grains.*—Why does the powder swell up when you heat it?—*The grains grow bigger.*—Are there more grains when we heat it?—*No, but the small grains grow bigger.*—And why does the flour swell?—*The heat is very strong.*—What does that do?—*It makes the small grains grow bigger.*—How?—...—Is there something that bothers you?—*Yes, I don't know how the small grains grow bigger.'*

BAS (12; 0). 'What has just happened?—*The heat has burst these yellow things* (the skin) *and so the maize has swollen.*—What do you mean by "swollen"?—*That the grains inside have grown bigger.*—What happens if we weigh them?—*The big ones weigh less, the heat has taken part of them away. Oh no, they stay the same, it's still the same weight.*—Why?—*Because nothing has gone, it's simply grown bigger.*—How?—*The heat has blown it up.*—What has it blown up?—*The grains.*—And if we counted them?—*It would be the same number. Here* (A) *they are smaller, and over there* (A') *they are bigger, but it's still the same number and the same weight.'*

The characteristic feature of these Stage III responses is therefore that the expansion of the whole seed is attributed to the 'swelling' of its elements, so that the weight no less than the quantity of matter remains constant.

The conservation of the weight, first of all, is explained in precisely the same way as it was in the case of the balls of clay (Chapter 2) and the sugar (Chapter 6). Alt, for example, put it all very clearly when he said : 'It's the same weight because none of the bits have dropped off.' This association of the weight with the substance is admittedly still questioned by the intermediate cases (Bert and Clav) who believe, at least at the beginning of the test, that when the individual flour particles 'swell up', their weight becomes changed, but by Sub-stage IIIB the conservation of the weight has been taken for granted, and the weight can henceforth be composed independently of the volume.

As for the 'swelling' schema by means of which these subjects explain the expansion of the total seed, it is the same we encountered at Stage II, except that it is now applied to the elementary particles of which the 'flour' or the powder is made up, and not merely to the seed as a whole. In fact, because he has learned to compose the weight as well as the substance, the child at this stage has taken a significant step towards atomism. At Stage II, it already happens that a subject may mention, not only skins or lumps of dough, but

126

also the elementary flour grains conceived as elements of the whole seed and even as being of constant number and substance. But because these subjects still deny the conservation of weight and make do with a global swelling schema, they fail to consider the behaviour of these small grains during the expansion. Subjects at Stage III may also contend that the grains are transformed into dough or lumps, but, more generally, they explicitly invoke a powder whose constituent particles are conserved during the general expansion resulting from the 'swelling' of the particles. Thus Bert explained that the number of 'small grains' remains constant, but that 'when it's opened up the grains become bigger'. According to Clav, the elementary 'parts' have 'grown bigger'. First they were 'all squashed together', but after the seed burst 'they're side by side but pulled apart'. Rou, for his part, believed that the 'small grains' remain constant in number when the seed bursts. Alt thought that 'the small grains grow bigger', because 'the heat is very strong'; though he added that he did not know precisely how, thus showing that he had given the matter some serious thought. According to Bas the grains grow bigger after the seed has burst, but 'it's still the same number'.

Atomism has thus become a quasi-general mode of explanation at Stage III. However, two quite distinct atomistic schemata can be used to explain the overall expansion of the seed. In the first, characteristic of Stage IV, the particles or 'small grains' remain at constant volume but draw apart when they are heated. This is the true operational schema of compression and decompression. In the second, characteristic of Stage III, the problem is simply shifted to the elements as such: the reason why the whole expands is that the atoms themselves swell up. It is clear that what we have here is not a real composition (except, of course, of the weight and the substance), but an intuitive transformation: the volume of each particle is thought to vary and the increase in global volume is attributed to an inexplicable increase in the total corpuscular volume. It is therefore no exaggeration to say that the intuitive 'swelling' schema of Stage II has been handed down as such to Stage III, where it is simply extended to the elements; and we must be careful to distinguish this schema from the operational schema of compression which does not appear until Stage IV. If we do so, we shall discover a precise parallel with the development described in Chapter 6 (dissolution of sugar), where the composition of the operations of compression also did not appear before the conservation of volume, and the latter was not constructed until after the conservation of weight. The only difference between these two developments, apart from the fact that the one involves dissolution and the other expansion, is that, in the second, the atomistic approach is less general than in the first: it can be replaced by a system of visible bits or parts. Thus while only a handful of subjects imagined the dissolved sugar as a kind of syrup, a relatively large number now believes that the maize seed is a kind of continuous dough or cottonwool that can be pulled out. However, even in their case, it is quite simple to distinguish the intuitive transformation of the volume characteristic of Stage III from the operational composition characteristic of Stage IV. In other words, atomism does not make a sudden appearance but is gradually consolidated from Stage II onwards, every stage having a characteristic mode of composition of which atomism is merely the culmination.

§3. *The fourth stage: conservation of the volume of
the particles and composition by compressions
and decompressions*

At Stage IV, the particles — no matter whether they are considered discontinuous grains or continuous lumps — are thought to be constant in substance, weight and volume, and the overall expansion of the seed is explained either by (1) the separation of the grains or (2) the inflation of the continuous lumps.

Let us begin with some responses of the first type:

LENS (9; 6). The weight will remain unchanged once the seed has burst. 'Why?—*It's the same amount of flour.*—Why has it grown bigger?—*It's blown out.*—Is the flour in one piece?—*It's full of little grains.*—Are there more or less of them when we heat them up?—*The same number.*—So why does it blow out?—*The small grains stay the same size, but they burst.*—What does that mean?—*They are less close, more apart.*'

CHEV (10; 0). The weight and substance will remain unchanged: 'Then why has it grown bigger?—*Because it's further apart.*—What is?—*What was inside, some white stuff: it's a small white powder.*—What is it made of?—*Small bits.*—What did you mean when you said "it's further apart"?—When we heat it, we make the bits smaller.*—How so?—*The bits separate out, they are further apart.*'

CHOL (11; 0). 'Is there more matter?—*No, it's no bigger, it's further apart.*—And if we put it on the scales?—*It will weigh the same, because it is a little bit emptier inside.*—What is the maize made up of?—*Of flour.*—Why does it separate when we heat it?—*It's full of small grains, which separate and burst.*'

JAC (12; 0). '*It's swollen up.*—What has?—*What was inside the seed, the flour, I mean the dough.*—Is there as much dough as before?—*Yes.*—And if we weigh it?—*It will be the same weight, only it's swollen.*—Why?—*Some air has got into the grain, hot air because of the heat.*—And then?—*Then the air gets into the grain of the flour and makes the big grains expand.*—Does the flour have grains?—*Yes, it's like a kind of powder.*—And when we heat it do we get a different number of grains?—*No, it'll be the same number.*—Will they get bigger or smaller?—*They'll stay the same.*—So why do they expand?—*Because the hot air pulls them apart.*—And how were these grains before the heat pulled them apart?—*Squashed tight together.*'

Here now an example of the second type (continuous structure):

SORO (11; 0). '*The heat has opened it up and then it burst.*—Why?—*The heat made it bigger.*—How?—*I don't know, I've never been told. There is some air inside.*—Are there more things in the seed than before?—*No, it's the same; it's just that the skin is liable to burst. It's grown bigger.*—How?—*It's opened out.*—What does that mean?—*It's collapsed, it's filled with air, by the heat I think.*—But how has it grown bigger?—*Here* (A) *it was all tight, that is squashed into a piece, but with the heat it's burst open.*—Does it weigh more?—*It weighs the same, because this one* (A') *is exactly the same as that one* (A).—How do you know?—*Because that one* (A) *is in a piece, it's squashed together, but this one* (A') *has burst with the heat.*—Draw the two seeds as you might see them under the microscope.' He draws a large circle for A, and a much larger one for A' with a set of intersecting lines representing the division of the grain. '*You can see that this one* (A) *is quite full, full of white matter. If we put that one* (A') *into this one* (A) *we would fill it completely.*—Are you sure they weigh the same?—*Yes.*—How do you explain that?—*In here* (the gaps between the lines in A') *there's nothing but air.*—How did it get in?—*It's got small holes, it's not all closed up like the first one.*—And what if we looked at them under the microscope?—*We should see a large ball with little holes.*'

These reactions differ markedly from those produced at Stage III: they show that the child has arrived at the most rational explanation of the expansion he can find. For the first time, he interprets changes in the global volume of the seed by applying the principles of conservation (including that of the total corpuscular volume) to its elements. In fact, unlike subjects at Stage III, he no longer believes in the 'swelling' of the 'small grains' or elementary particles, i.e. he no longer explains the global expansion by shifting the problem. Lens and Jac, for example, both declared that the small grains 'stay the same size'. As for Soro, who thought that the dough 'opens out' during the expansion, he too did not speak of a 'swelling' or change in concentration of the particles. In other words, all these children have begun to realize that the increase in the volume of the whole seed is the result of a spatial composition of elements kept at constant volume. According to Soro, the seed 'bursts with the heat' but is 'squashed together' in the normal state, i.e. the seed is 'in a piece' or 'quite full' when it has not been heated, but bursts when the hot air gets in through the 'small holes'. According to Lens the 'small grains' burst and draw apart. According to Chev the 'small bits' inside the seed become smaller and the resulting powder grains 'separate out'; and according to Chol the 'small grains separate and burst'. Jac produced a very precise schema when he said that the grains, which are normally 'squashed tight together', do not change as such when the seed bursts, but that 'the hot air pulls them apart'.

It is clear that this schema, in which the compression and decompression of the maize seed is attributed to its division into grains or particles and their subsequent displacement, is the result of the same grouping of operations we met in Stage IV responses to the dissolution of sugar cubes (Chapter 6) but which has now become explicit. The discovery of these operations is reflected particularly well in the invention of the inverse operations. At earlier stages, the 'swelling' was considered an irreversible transformation and was therefore clearly distinct from the operational schema of decompression: even if the subject might have imagined that the 'swollen' grains could be restored to their initial state, he would have been thinking of an empirical process, not an inverse operation, and this precisely because he was still unable to compose the total volume of A' with the volumes of the elements in A. By contrast, when Soro explained: 'If we put that one (A') into this one (A) we would fill it (A) completely', he was mentally performing the operation of compression and its inverse and hence was able to prove the correctness of his explanation. The operational reversibility of the relations is thus affirmed for the first time and this advance is, in fact, the chief characteristic of Stage IV.

At Stage I, irreversibility is complete because the expansion of the seed is attributed to a process of biological growth, i.e. to a one-way development. At Stage II, we have the first type of reversible composition, which is confined to the substance: the initial seed is thought to consist of a skin, pieces of dough, or small grains of powder, etc., and each of these grains can be rediscovered in more or less changed form in the expanded seed, which is believed to have changed in weight and volume but not in the quantity of matter. Thus if S is the total initial substance and s_1; s_2; s_3 ... are its parts; and if S' is the final substance, and s'_1; s'_2; s'_3 ... etc. its parts, then we have $s_1 + s_2 + s_3 ... = S$, and $s'_1 + s'_2 + s'_3 ... = S'$. Moreover, $s_1 = s'_1$; $s_2 = s'_2$; $s_3 = s'_3$ etc., whence $S = S'$, no matter what the apparent differences between them. But neither the weight nor the volume of S and S' can yet be composed in this or in any other reversible way. In fact, the child at Stage II imagines the expansion of the seed as a kind of global swelling under the effects of the hot air, which radically transforms the entire system in an irreversible manner. For example, let us suppose that those children who believe in a simultaneous increase in the weight and the volume (i.e. that $V' > V$ and $W' > W$, where V and W are the original volume and weight and V' and W' the final volume and weight), make use of the composition $V' = V + V''$, where V'' is the increase in volume due to heating, and $W' = W + W''$, where W'' is the increase in weight due to the expansion in volume. Hence either V'' is simply the volume of the hot air, i.e. of the interstitial spaces between the parts of V, and we would have a Stage IV response, which is not the case; or else V'' is nothing but the empirical difference, in which case it is possible to invert the expression and to apply the formula $V = V' - V''$ for the purpose of classifying or seriating the data. But it is not yet possible to construct V' or V'' with the elements of V because these elements themselves are thought to change during the expansion. Thus if the volume V is composed of the partial volumes $v_1 + v_2 + v_3 \ldots$ and if $V' = v'_1 + v'_2 + v'_3 \ldots$ we no longer have $v_1 = v'_1$; $v_2 = v'_2$, etc. The same difficulty impedes the composition of the weights.

But even if the child assumes that the weight decreases with increases in volume, can we really claim that he has grasped the reversible composition of the ratio W/V? It goes without saying that by distinguishing the relative weight W/V from the absolute weight $W = w_1 + w_2 = w_3 \ldots$ he could arrive at this composition, but at Stage II he still confuses W with W/V so that he is once again unable to construct W' from W; in the absence of an invariant element, the swelling and associated relations cannot possibly constitute a reversible grouping. Admittedly, the child may seriate what differences he perceives or posits subjectively, classify their terms, or even seriate and classify them simultaneously, which comes back to counting or measuring, but the compositions resulting from these formal operations are no more than logical groupings or arithmetical groups, i.e. systems of relations or laws (which may be true or false). But causal explanations cannot appear until the very content of these relations (the classified, seriated or numbered objects) is seen to be invariant, i.e. until reality itself has been grouped, which is not possible without physical operations of a spatio-temporal kind.

Moreover, as we shall see in Part IV, formal relations as such cannot be constructed until their content is available for composition. Thus when the child finds that A has the same weight as B, and B the same weight as C, he does not necessarily conclude that $A = C$ from $A = B$ and $B = C$; to do so he must first treat the weight as a quantum. And if $C = D$, he does not necessarily conclude that $(A + B) = (C + D)$; indeed, if A is made of lead and B, C and D of iron, he generally considers $(A + B)$ heavier than $(C + D)$ even after discovering the equalities of $A = B = C = D$, on the grounds that 'iron and lead are heavier than iron by itself'. Now, when subjects at Stage II assume that W' (the weight of the expanded maize seed) is smaller than W (the initial weight) because the additional W'' (the weight of the hot air) is light they argue in precisely the same way. Thus, as Mey explained, the 'open' seed weighs less 'because there's more air inside ... it's the same (inside) but it's lighter'; or as Mat put it: 'The air weighs less than the seed; it takes away part of the weight'; i.e. $W' < W$ because $W' = W + W'''$! In other words, at Stage II the composition of weights and volumes remains irreversible, so that the atomism characteristic of this stage does not yet constitute an operational schema, except in respect of the quantity of matter; the elements $s_1, s_2, s_3 \ldots$ of A and A' are seen to be identical in substance but do not yet constitute units or even logical identities of weight or volume.

During Stage III reversibility is extended from the substance to the weight; i.e. the child comes to realize that, despite the expansion of the maize seed, the particles $s_1 + s_2 + s_3 \ldots = S$, which are conserved in S' in the form of $s'_1 = s_1$, $s'_2 = s_2$, etc. also conserve their weight, whence $(W = w_1 + w_2 + w_3 \ldots) = (W' = w'_1 + w'_2 + w'_3 \ldots)$. But the method by which the child explains the expansion remains irreversible, because he still treats the latter as a 'swelling', this time of the elementary grains themselves. Now, no composition is possible if the child posits $v'_1 > v_1$, $v'_2 > v_2$, etc. to explain that $V' > V$, thus shifting the problem from the whole to the parts.

Complete reversibility is finally reached at Stage IV, where the schema of compression and decompression replaces the 'swelling' schema. Thanks to this new system of composition, the child explains the volume V' of A' exclusively by reference to the elements of V and their displacements, and treats the compressions and decompressions as centripetal or centrifugal rather than random displacements. As a result, the elements of the maize seed are seen to be of constant volume as well as of constant substance and weight, and the child can distinguish between the constant total corpuscular volume $(V = v_1 + v_2 = v_3 \ldots)$, and the variable global volume, so that atomism acquires its characteristic operational coherence and can give rise to causal explanations. Atomism, in fact, is nothing but the expression of the grouping that leads the subject to, and is completed with, the construction of the conservation of substance, weight and volume; and this precisely because the latter define the compositional elements to which we refer as atoms, and thanks to which it is possible not only to explain the invariances revealed by the experiments but also such apparently mysterious changes as the expansion of the maize seed after it has burst open. Hence, when we compare the results of the maize seed test with those obtained in the sugar and clay experiments, it is not surprising that we should discover so many close parallels in the child's gradual advance from simple conservation to atomism and again to the schema of compression and decompression.

§4. *The expansion and contraction of mercury*

We also tested the responses of our subjects to the fluctuation of a column of mercury in a thermometer tube. Since the mercury suggests the idea of drops or visible parts rather than of invisible atoms, it provides an excellent method of verifying our other conclusions. Now, as we shall see, though this test produced poorer representations, the responses nevertheless had the same logical rigour as the last, which is all that concerns us here.

To the least advanced subjects (Stage I) the expanding mercury conserves neither its weight nor its quantity of matter; the expansion is treated as a simple increase in substance:

HEN (7; 0) cannot tell why the column rises: 'Is there more liquid, more matter, than before or not?—*A little bit more.*—Are you just saying so or is there really some more?—*There is more liquid.*—And if we weigh it?—*It would be a little heavier.*—And when it drops back again?—*There'll be less inside and that will be lighter.'*

VOL (9; 0). *'It rose because it was hotter.*—Why does it rise up when it gets hotter?—...—And what if we weigh it?—*It'll be heavier.*—Why?—*Because we added some more.*—Some more what?—*I don't know.*—Are you only saying that something has been added or is it really true?—*No, it's true. It's heavier because there's more mercury.'*

Subjects at Stage II acknowledge the conservation of substance, but continue to believe that the weight varies, either positively or negatively.

EPA (8; 0). When you blow on the tube *'the liquid goes up.*—Why?—*Because.*—What?—*Because it's warm.*—Will there be more liquid or will it be the same?—*It'll be the same amount, but it'll rise higher.*—Why?—...—And the weight?—*The liquid is a little lighter, because there is a little less of it in the same place* (= because it expands).'

MADEN (10; 0). *'When it's cold, it's like a stone, but the heat turns it into a liquid and pushes it up.*—How?—*When it gets hot, it makes steam, and the steam pushes it.*—Is there more liquid than before or not?—*No new liquid has been made, but it takes up more room.*—And the weight?—*It weighs more when it's up.'*

ZUM (11; 0), before being asked about the maize: *'The heat expands the mercury; it makes it swell up when it gets in. The heat takes up room and the mercury swells.*—How does the heat get in?—*It's very fine, it's liquid*—What if we put the thermometer into cold water?—*When it's cold it shrinks, and when the heat gets in it gets bigger.*—Why does it shrink when it's cold?—*It gets a little smaller because it contracts.*—Is it the same weight when it expands?—*The heat takes up room and it's lighter, so the whole thing becomes lighter.*—And when we cool it down?—*There's no more heat, so it gets smaller and becomes the same weight as before.'*

There is thus a close analogy between these Stage II reactions by children none of whom had taken the maize test, and those who had. To some, like Maden, the expanded mercury weighs more because it is more voluminous (which is the more surprising as he also explained that no fresh substance had been added). To others (and these were in the majority), the expanded mercury weighs less. Epa put forward a purely subjectivist view when he said: 'The liquid is a little lighter, because there is a little less of it in the same place.' As for Zum, he provided a good example of empirical rather than rational reversibility (see §3) when he claimed that the heat caused the mercury to 'swell up' as it 'gets in' and so made it lighter and that, conversely, when 'there's no more heat' the mercury 'gets smaller and becomes the same weight as before'. He thus made it perfectly clear that he was not applying true reversibility: if W is the initial weight of the mercury and W'' the weight of the 'heat substance', then W' (the weight of the expanded mercury) = W + W''; but, according to Zum, 'the heat takes up room and it's lighter, so everything becomes lighter'! In other words, he failed to compose the weight by means of an additive grouping, and relied instead on an egocentric evaluation that might be expressed as 'a heavy object plus a light object weigh less than the heavy object alone'. We shall see in Chapter 10 that this approach is typical of subjects who have not yet advanced to groupings.

During Stage III, there is complete conservation of the substance and the weight of the mercury (as of the maize), but the expansion is attributed to the 'swelling' of the elements considered as separate drops or as simple parts of a continuous whole:

MOS (11; 6). *'It rises up, but when it's cold it will contract.*—Why?—*Because when it's hot, the droplets stretch out and rise.*—What droplets?—*When you spill mercury it forms little balls.*—And when it's cold?—*They'll drop back to the bottom.*—Is it heavier when it rises?—*No, it isn't. The small balls haven't changed, it's just that the heat has made them stretch out and expand.'*

ROG (11; 10). *'It rises, it increases with the heat.*—Is there more or less mercury than before?—*The same. When it's hot it melts, it gets runny, but there's no more of it.*—And the weight?—*It gets heavier. Oh, no, it stays the same because it's as much mercury.*—But why does it rise?—*It takes up more room, like water when it freezes.*—Does ice weigh more than water?—*No, the same.*—Why?—*Because nothing new has been added.'*

It is interesting that, though they do not use an atomistic approach, these children employ precisely the same constructions as their Stage III counterparts do with the maize seed. Here, now, are a few Stage IV responses in which the empirical schema has made way for a (more or less advanced) schema of spatial composition:

VEL (11; 0). 'Why does the mercury rise?—*When the mercury is cold it's like a piece of ice, and when it's hot it stretches.*—Is it hard when it's cold?—*No, when it's cold it is more squashed and when it's hot it's more stretched out.*—Is it squashed now?—*Yes, when it's cold it's crammed together like children in a gym touching shoulders.*—What is crammed together?—*The parts of the mercury.*—Are there more of them?—*No, they always stay the same.*—Why?—*If you break a thermometer at nought degrees there'll be a certain mass of mercury, and if you break it at eighty degrees there'll still be as much.*—And the weight?—*It's the same weight because it's just stretched out.*—How so?—*When it's hot it tends to evaporate, so it rises up. Then it splits up into little bits as it rises, but it keeps the same weight.'*

CIE (12; 0). 'When it's hot it makes the substance still more liquid.—Why?—*If you heat a bit of chocolate it gets runny and takes up more room.*—Why?—*Before it was hard, now it's expanded.*—How so?—*The parts of the liquid are all squashed together, but when you heat them they spread out.*—And the weight?—*It's the same, the same parts.'*

LIC (13; 0). 'It's the heat that makes it expand.—Is there more mercury?—*No, it's the same amount but now it's stretched out.*—How so?—*It gets thinner, less together.*—What does that mean?—*It's less squashed.*—How so?—*It's the hot air, it mixes with the mercury. When you heat it you get bubbles of hot air in the mercury, and that pulls it apart. No, not bubbles of air, it's bubbles of mercury; tiny little balls that draw apart when there's air inside them.*—And when it's cold?—*They are more together, they are all piled on top of each other.'*

We see that, though these children do not adopt the idea of a granular structure, which the maize flour or powder suggests to their counterparts quite naturally, the very logic of the arguments by which they construct their invariants nevertheless leads them to the schema of the spatial composition of the parts. In fact, when children at Stage III first grasp the fact that the expansion of the mercury changes neither its quantity of matter nor its weight, they already think in terms of a sort of stretching due to the 'swelling' of the parts, but simply shift the problem to the latter. But once they have come to appreciate this difficulty, they are forced to apply the same schema of composition to the volume that has enabled them to construct the conservation of weight and substance despite all apparent changes, with only this difference, that the volume varies in fact. They then imagine the existence of 'parts' or 'balls' whose total volume is constant but whose global volume varies with the arrangement, and hence apply the schema of decompression to the expansion of the mercury as such. According to Vel, for example, the expansion presupposes a division: 'When it's hot it tends to evaporate ... it splits up into little bits as it rises'; but when it is cold 'it's like a piece of ice ... the parts of the mercury' are 'crammed together like children in a gym touching shoulders'. Cie, too, contended that, when they are cold, 'the parts of the liquid are all squashed together, but when you heat them, they spread out'. Lic described the mechanism responsible for the decompression quite explicitly when he said that, when the mercury is cold, the 'balls' of which it is made up are 'all piled on top of each other'; but when it is hot they are less 'squashed' because the air has got between them. His explanation was therefore identical with Stage IV explanations of the expansion of the maize seed.

In other words, even in this series of tests which lend themselves so badly to an atomistic interpretation, we witness the construction of the same invariants with the help of the same compositions. This shows once again that the schema of compression is constructed as part of an overall system.

CHAPTER EIGHT

Differences in Density

Having examined the construction and atomistic composition of the schema of compression and decompression, we must now try to determine to what extent this schema helps the child to explain differences in density. To that end we show him a cork and a smaller but heavier stone (or a piece of wood heavier than the cork, lighter than the stone and of intermediate volume), and ask him which is lighter, which is heavier, and why. We also produce two stones of the same shape and size but of different weights, e.g. a pumice stone and a pebble (or a silver and a brass franc), and ask for an explanation of the differences. Now, the combined answers we obtained could be fitted more or less clearly into the four stages of development we have distinguished in previous chapters. During Stage I the child fails to dissociate the weight from both the volume and the quantity of matter and believes that the larger a body the heavier it must be; and once the balance shows him that this is not necessarily the case, he explains the results by intuitive arguments based on the different origins or methods of growth of the two objects. During Stage II the weight becomes dissociated from the apparent quantity of matter, but the differences in density are still explained by intuitive arguments and there is no composition of the volume or the weight of the elements. At Stage III the differences in weight at equal volume and the fact that the weight is in inverse ratio to the volume are explained in terms of the quantity of matter and the weight of the elements composing the two bodies, but the resulting composition does not yet involve compression and decompression. Finally, at Stage IV, differences in density are attributed to the compression or decompression of the elements (spatial composition).

§1. *The density of a cork, a piece of wood and two pebbles: Stages I and II*

Here, first of all, are four Stage I responses to the pebble and cork test:

KEC (4; 6) says that the cork is heavier than the pebble *'because the pebble is smaller.*—Are smaller things always lighter?—*Yes, when things are big, there's more inside them.*—And why do you think that this cork is heavier than the pebble?—*Because there are more things* (= substance) *inside.*—See for yourself.' He weighs them in his hand. 'Which is heavier?—*The cork is lighter.*—Why?—*Because there's nothing inside but cork, and cork isn't heavy.*—And what about this piece of wood? Is it lighter or heavier than the pebble?—*The pebble is heavier.*—Why?—*Because wood is lighter, and it's smaller as well* (the piece of wood was, in fact, larger than the pebble).—But is it bigger than the pebble?—*Yes.*—Is it heavier or lighter?—*I don't know.'*

DAD (5; 0). *'The cork is heavier.*—Why?—*It has to be, because big things are heavy.*—Weigh it in your hand (he does so). Well?—*The pebble is heavier.*—Why?—*Because that's how it is.*—And what about this piece of wood?—*The stone will be heavier.*—Why?—*Because stone is for making roads and wood is for making tables.'*

GUER (6; 0). 'Which is heavier?—*The cork.*—Why?—*Because it's thick.*—See for yourself.' He weighs it in his hand. *'It's the pebble.*—Why?—*Because the cork is made of cork, and the pebble is a stone.*—Why is a stone heavier?—*Because corks are for bottles and if they were heavy the bottles would break.*—But why is the stone heavier, even though it's smaller?—*Because houses are built of stone.'*

ZUR (8; 1). 'Which is heavier?—*The cork.*—Why?—*Because it's bigger.'* He weighs it in his hand. *'No, it's the stone.*—Why?—*Because the stone is bigger than the cork, and so it must be heavier.*—But is it really bigger?—*No.*—So?—* . . . —Why is it heavier?—*The stone is smaller! If it's smaller it must be heavier!'*

Eight days later we saw Zur again and asked him the same questions: *'The stone was heavier.*—Why?—* . . . —What is bothering you?—*The cork was bigger!*—Here, take two boxes of matches. If this one is heavier than that one, why is that?—*Because there are more matches inside.*—And why is the stone heavier than this cork?—*Because it's white inside.'*

Next for some Stage I responses to the presentation of two pebbles or coins of the same dimensions:

BAR (5; 0) thinks that the brass franc is lighter than the silver franc *'because it's yellow'.*

CHE (6; 0) thinks that the brass franc will be heavier *'because it's thicker.*—See for yourself.' He weighs it in his hand. *'No, it's that one.*—Why?—*Because that one is thicker.*—What does that mean?—*That it doesn't weigh the same.'*

FRA (6; 6). 'Which is heavier?—*That one.*—Why?—*It's bigger.*—Bigger than the other?—*No.*—So why is it heavier?—*Because.'*

BER (6; 0). The pebble is heavier than the pumice stone *'because it's a flint, you can light matches with it'.*

LIR (7; 0). The silver coin is heavier *'because the yellow one has been made lighter.*—But how did they make it lighter?—*Because they made it like that.'*

DRA (7; 0). The silver coin is heavier *'because it's of white metal.*—But why is white metal heavier than yellow?—*Because that's the kind of metal it is.'*

MA (7; 6). The pebble is heavier than the pumice stone because *'stones are made. They grow. At first they are very small, like small stones, and before that there was nothing.'*

SI (8; 0). *'The grey stone is lighter because it comes from the sea and the other from the lake. There's more water in the sea, and so there are more stones: they're lighter, it's always been like that.'*

We see that these Stage I responses correspond to those we have discussed in earlier chapters. There is no distinction between substance, weight and volume; there is no conservation; and there are no atomistic schemata transcending the perceptive data. In fact, all these children imagine that weight is proportional to length or thickness, and hence to both the quantity of matter and the global volume. Thus Kec thought the cork heavier than the pebble, because 'when things are big, there is more inside them'. Dad, Guer and Zur explained that 'big things are heavy', and that the pebble is heavier 'because it's thick' and 'because it's bigger'. Now since objects of equal density increase in weight as they increase in volume, these responses might simply suggest that four- to seven-year-olds are unaware of the low density of the cork. In fact, however, not only have most of these children handled corks at home, but they invariably tell us that cork is a light substance. Thus Kec said: 'There's nothing inside but cork, and cork isn't heavy'; and Guer explained that the cork was light 'because it is made of cork'. Hence it was not their ignorance of the relative weight of cork that made these children declare that the pebble was lighter: the real reason was that they could not dissociate the quantity of matter from the volume and the volume from the weight, or more precisely that they failed to grasp that A could be both heavier and smaller than B. When the two objects are of equal volume, these children find it easy to predict their specific weight or density, but as soon as the weight and the volume are in inverse ratio they consider the latter proportional to the former, so much so that, even after weighing the stone and the cork, they still insist on retaining the schema weight = volume = quantity of matter, thus denying the evidence of their own senses. Kec, for instance, claimed the wood was smaller than the pebble when, in fact, it was larger, and Zur, though considerably older, went further still: when he realized that the pebble was heavier though smaller than the cork, he invoked what we have elsewhere[1] described as logical participation and explained that 'the stone is bigger than the cork, and so it must be heavier'. Eight days later he was still wondering why this was so, for he remembered that the stone was, in fact, the smaller of the two

[1] See J. Piaget and A. Szeminska, *The Child's Conception of Number*, Routledge & Kegan Paul, 1952, Chapter VII.

objects he had been shown. Similarly, Che declared that the brass franc was heavier than the silver coin because it was 'thicker', but seeing that the silver coin weighed more he ended up by declaring that *it* must be the 'thicker' of the two! Again, Fra contended that the 'yellow' coin was heavier because it was bigger.

In short, at Stage I there is no quantitative differentiation of the weight from the volume and the quantity of matter. However, since the child cannot help noticing differences in weight as soon as he lifts up the different objects, he is forced to seek an explanation, but fails to reach a quantitative correlation of weight and volume because he has been in the habit of confusing what has suddenly proved to be quite distinct. The simplest solution is to turn both weight and volume into substantial qualities and to leave it at that. Hence when he tries to account for his own observations, he may simply argue, like Kec, that the cork is lighter because 'there's nothing inside but cork'; like Guer, that the cork is lighter because 'it is made of cork, and the pebble is a stone'; and like Dra, that the silver coin is heavier 'because that's the kind of metal it is'. Others again attribute the weight of an object to some inherent quality by a purely phenomenalistic association – the brass franc is lighter 'because it is yellow' (Bar); or the pebble is heavier 'because it's white inside' (Zur). Or else they adduce some sort of final cause: the pebble is heavier because 'it has to be, that's how it is' or because it is 'for making roads' (Dad). Similarly Guer thought that the cork must be lighter since otherwise it would break bottles and the stone heavier because it is used for building houses. From finalism the child thus often proceeds to artificialism, or else he may invoke the place of origin or process of growth of the object under discussion: 'Stones are made,' Ma explained. 'They grow. At first they are very small, and before that there was nothing.'

Stage II brings a marked advance: the dissociation of the weight, the quantity of matter and the volume. In the previous chapters, we saw that the primitive stage of complete non-differentiation is always followed by a second stage during which the conservation of matter (but not yet of weight or volume) is constructed; and an elementary form of atomism not yet based on a genuine spatial composition serves to explain the permanence of the distorted, or dissolved, expanded or contracted substance. Now the same stage also appears in the present series of experiments, where it takes the form of the correct evaluation of weights based on differences in density, i.e. on the discovery that the weight of an object does not solely depend on its volume but also on 'what is inside'. However, at this stage, the weight is not yet sufficiently quantified for its conservation to be extended to the case of changes in shape (*cf.*

Chapters 2, 5 and 7); weight remains an intuitive notion of a pheno-menalistic and egocentric kind. In other words, it remains a substantial quality, and the only advance of subjects at this stage is that they try to relate it to the contents of the object and no longer solely to its volume. However, this new perspective does not yet lead them to compositions capable of explaining the real relationship between the weight and the quantity of matter, or between the weight and the volume. Hence differences of density are still treated in much the same way as they were at Stage I.

DUF (7; 0). Before producing the test objects, the demonstrator asks: 'Are big things always heavier than small ones?—*No.*—Why not?—*Because some big things are lighter than small ones.*' A pebble, a piece of wood and a cork are brought out: 'Which is heaviest?—*The pebble.*—And lightest?—*The cork.*—Why is the pebble heavier than the cork?—*Because one is made of cork and the other of stone.*—And why is stone heavier than cork?—...' (In the clay ball experiment, Duf had also adopted the conservation of substance but not of weight.)

JEAD (8; 0). The pebble is *'smaller because it's a small stone, but it's heavier because it's made of stone.*—Why?—*Because it's made of earth.'*

ROY (8; 0). Three objects: 'Which is heaviest?—*The stone.*—And lightest?—*The cork.*—Why?—*Because it's cork.*—And which is biggest?—*The cork.*—And smallest?—*The stone.*—So why is it heavier?—*Because it's a pebble.*—Why does that make it heavier?—*It's got more weight.*—And why is the cork lighter?—*Because what's inside is lighter than what's inside the pebble.*—What exactly is inside?—*Wood from a tree.*—Why is that lighter?—...—Which is heavier, this cork or this piece of wood?—*The wood.*—Why?—*Because it's not from the same tree.*—So?—*It's got more weight.*—But why should it be heavier when it's smaller?—*Because it's not the same wood. Here* (he points to the cracks in the cork) *there are some places with no cork, holes you might say; they're not properly filled.'* The demonstrator splits the cork. *'Oh, no, it's quite full.*—So?—*The cork weighs less because it's not made of the same tree.'*

CHA (9; 0) is another subject who was found to be at Stage II with the clay ball test: when asked to reduce the weight of one of the balls, he merely squashed it as hard as he could. When shown the pebble and cork, he says: *'The pebble is heavier, because it's made of stone.*—Which is bigger?—*The cork.*—So?—*The pebble is heavier because it's smaller, it's made of stone.'*

Now for some Stage II comparisons of two stones:

LOU (8; 8). *'This one* (the pumice stone) *is lighter because it comes from the bottom of the sea.*—But why does that make it lighter?—*The water makes it softer and lighter. You can see that the other one comes from the mountains.'*

JEA (9; 6). *'The water changes it. This one* (the pebble) *gets heavier in the water.*—How so?—*Water is heavy. When a stone is put in the water it gets heavy as well, and when the other one* (the pumice stone) *is not in the water, it gets lighter.'*

EL (10; 0). *'It* (the pumice stone) *is a young stone; it's lighter because it hasn't been planted; it hasn't stayed so long in the ground, and so it's less cultivated, less heavy.'*

GAR (10; 0). *'These stones come from different places and so they are either heavier or lighter.—*Where do these stones come from?*—From all over the place. The light one comes from where the earth weighs less.—*Why?*—In the mountains it's drier. In the water the earth weighs more when you take it out, because of the weight of the water. Mountain stones are lighter, they come from dry places.—*What if we carried the heavy stone up into the mountains?*— It'll still be heavier because it's grown in the water.'*

A comparison of these Stage II responses with those we have met previously is highly instructive.

To begin with, all the subjects who declared straightaway that the pebble was heavier than the cork and the piece of wood, had been questioned about the clay ball, and all had assumed the conservation of substance but not that of weight and volume. Now the fact that they consider the respective weights of like objects inversely proportional to their volumes and that they even predict that the larger of two distinct objects will not necessarily be the heavier, shows that in the present series of tests, too, they no longer confuse the weight with the total quantity of matter or with the volume. Just as they believe that, when the shape of the clay ball is changed, or when the maize seed bursts, the substance is conserved but the weight increases or decreases, so they also believe that the cork is larger than the pebble, and hence contains more substance, but is nevertheless lighter because of the quality of that substance.

But what precisely are we to make of this peculiar conception of weight and, more particularly, how do these subjects explain differences in density? It is worth dwelling on this question because nothing throws more light on progress in the composition of physical operations than the contrast between their initial and final explanations of differences in density.

Most subjects at Stage II believe that the quantity of matter is directly proportional to the volume, and inversely proportional to the weight. Subsequently, however, and thanks largely to the concept of relative 'fullness' (Stage III) and relative 'tightness' (the schema of compression characteristic of Stage IV) they re-establish direct proportionality between quantity of matter, volume and the weight, but distinguish the corpuscular quantity of matter from the apparent or global quantity, and the corpuscular volume from the global volume. As a result, they come to assume that the pebble contains a greater number of elements than the cork, but in a compressed state, even though the cork, being the larger of the two, seems to contain more matter. Now it is this general composition of a corpuscular type that then enables them to explain the differences in density, while maintaining simple operational relations between the substance (which is thus almost promoted to the status of 'mass'), the weight and the volume thanks to the schema of compression and decompression. In short, much as the expansion of the maize seed is eventually treated as a change in global volume, which leaves the substance, the weight and even the corpuscular volume unchanged, so the differences in the density of the pebble, the piece of wood and the cork are ultimately attributed to the fact that the weight varies inversely as the apparent quantity of matter and the global volume, and directly as the real quantity of matter and the corpuscular volume. It follows that the mistake of subjects at Stage I is simply that they apply direct proportionality to the apparent data, instead of distinguishing the latter from the corpuscular relations. Once this is corrected, the initial relations become operational and lend themselves to reversible composition.

Now, while subjects at Stage II believe that the apparent or global quantity of matter varies more or less as the global volume, they cannot, for lack of an adequate corpuscular composition, distinguish these totalities from the quantity of matter or volume of the elements. Quite plainly, children at this stage are still quite incapable of conceiving of a body as an invariant whole, made up of substantial parts of constant weight or volume. In particular, since they still fail to consider the relative compactness of the parts, they cannot but consider the total quantity of matter as being directly proportional to the global volume, even if the latter varies during deformations while the first remains constant. But in that case, how do they account for differences in density?

Since, at Stage II, they cannot do so by reference to the volume or the total quantity of matter, they have no alternative but to rely on intuition – which, as we saw, is also a typical Stage II approach to differences in weight. This explains why our present subjects, like those at Stage I, still invoke substantial qualities based on the mode or place of origin of the various bodies, their only progress being that they have moderated their animism and artificialism and try to explain differences in weight by purely qualitative combinations. Thus Jead said that the stone was heavier because it was made of earth, and Roy invoked 'what is inside' and, making no attempt at composition, concluded that the different 'woods' had different weights. He might, however, have seemed to be on the right track when he contended that the cork was not 'properly filled'; but, in fact, he did not apply this phrase to the corpuscular structure; what he had in mind was rather a series of holes (it was the cracks in the cork that had given him this idea), and he recanted

143

as soon as he saw the inside of the split cork. With the two stones, the most common argument is that the water has made the pumice stone lighter, which explains the greater weight of the pebble ... or the precise opposite (*cf.* Lou and Jea). According to El, the pumice stone was lighter because it was young and the pebble heavier because it had stayed longer in the ground: 'it's less cultivated'. Gar explained that the original qualities are always conserved even during displacements: if the pumice stone were transported to a mountain, 'it'll still be heavier because it's grown in the water'. It is curious that these children, who do not grasp the conservation of weight during the deformations of a solid, should nevertheless believe that the original, imaginary, qualities are so persistent – at least when their arguments demand that they should be so!

§2. *Stage III : explanation of density by relative fullness*

At Stage III there appear two successive explanations of density by corpuscular composition: one corresponding to the notion of relative fullness and the other based on the schema of compression and decompression. The difference between the two is similar to that between the distension of the elements of the maize seed and the schema of their relative compression. It is a characteristic of Stage III that the conservation of weight becomes operationally linked to the quantity of matter. However, since the weights of the pebble, the piece of wood and the cork are inversely proportional to their overall volumes, and hence also seem inversely proportional to their quantity of matter, the only apparent means of co-ordinating the weight and the substance is to consider the latter as more or less full, and this is precisely what our subjects do, though by means of a quite novel distinction, namely that between the apparent or global quantity of matter and the corpuscular or total quantity. In what has gone before, we have largely been dealing with like objects that conserve their substance during transformations, while in the present chapter we are dealing with objects of different densities. Thus, even with the maize seed, the child could conceive of a swelling of elementary granules, which conserve their substance, whereas, in the present case, he must attribute the relative fullness of bodies to quantitative differences in corpuscular matter.

In analysing his conception of the volumes of the pebble, cork, etc. we must, in fact, be careful to distinguish between 'fullness' and 'tightness'. When he says that something is 'full', he is thinking of boxes filled, say, with matches: the fewer matches there are, the lighter the box. In other words, there is no real spatial composition, and often no precise atomic structure, but merely the idea of *plena* and *vacua.* 'Tightness', by contrast, i.e. the schema of compression and decompression, implies the idea of corpuscles whose separation is subject to a spatial composition, the global volume being inversely proportional to the tightness. Needless to say, we shall discover a host of intermediate reactions, especially when, lacking rigorous arguments, a subject renounces one approach in favour of the other.

Here, first of all, are some responses by subjects who are on the brink of Stage III (Sub-stage IIIA):

MAZ (7; 6) believes in the conservation of matter, but not yet in that of weight. He immediately produces correct assessments of the weight of the pebble, the piece of wood and the cork, having first stated that big objects are not always the heaviest: 'Why not?—*Because the pebble is heavier than the piece of wood,*' even though the wood is '*bigger.*—How is it that the pebble is heavier?—*Because it's fuller.*—What does that mean?—*That it's thicker.*—Is it bigger than the wood?—*No, it's smaller.*—Then how can it be thicker?—*Because the small pebbles are thicker than the bits of wood.*—Why?—*Because they are fuller.*—And which of these two (cork and wood) is heavier?—*That one* (the wood).—And thicker?—*That one* (the cork). Why?—*Because the cork is bigger than the wood.*—Then why is it lighter?—*Because sometimes bits of wood are heavier than a cork, because cork is lighter.*—But how is it that wood is both smaller and heavier?—*Because there are more things in the piece of wood and fewer things in the cork.*— What does that mean?—*That it's fuller.*'

BAE (9; 6) accepts the conservation of matter but is not very sure of the conservation of weight: '*The stone is heavier because that stopper is made of cork.*—Why is the stone heavier?—*Because the stone is made of hard stuff.*—Why does that make it heavier?—*Because it's different inside: the cork has lots of holes and the stone is full of sand.*—Why does that make it heavier?—*Because it's stone.*—And what about this one (a ball of modelling clay the same volume as the stone)?—*The stone is heavier because the Plasticine is made with earth. It's got water inside and water is liquid.*—And so?—*The Plasticine is lighter because it's not so hard.*'

DEBO (10; 0) also begins by associating weight with hardness, but then advances to the schema characteristic of Stage III: '*The wood is the biggest and the pebble the heaviest.*—Why?—*Because the piece of wood is longer and the stone is harder.*—Why is it heavier when it's harder?—*Because it's full inside.*—And the wood?—*It's light because it's long.*—How so?—*It is long, so it can stretch more easily* (= it is not so hard).—And so?—*It's stretched, it's lighter because it has more size* (*cf.* the concept of "swelling").'

Next for some clear Stage III reactions (Sub-stage IIIB):

HUB (8; 0) assumes the conservation of weight. Wood and cork: '*The wood is heavier.*—Why?—*It's fuller.*—And that one (the stone)?—*The stone is the heaviest of all; it's fuller still.*—Why?—*It's chock-full.*'

HAL (9; 0). '*The cork is not so heavy, and the stone is heavier.*—Why?— *There are more things inside and that makes it heavy.*'

SCHO (9; 0). '*The stone is heavier because it's full, there's more sand inside.*'

145

REYB (10; 0). *'The pebble is fuller than the cork.'*

KRE (10; 0). *'The stone is full; the cork is more puffed up, it's not so full.'*

GRO (11; 0). *'The pebble is full and the cork is less full; we can stretch it out (cf.* the "swelling" schema).'

PAT (12; 0). *'The stone has more things inside; it's fuller, and that makes it heavier.'*

And now for some comparisons of the two stones:

ALB (8; 8). 'Are they as heavy as each other?—*It's hard to say* (he feels them in his hand). *Oh gosh, just look at it* (points to the balance), *you could say it* (the pumice stone) *was completely empty.*—So?—*This one* (the pebble) *is all full, and that one is all empty.'*

OLT (10; 0). *'One is a little bit empty inside; there may be little holes in it. The other is all full.'*

LIL (11; 7). *'It's because some stones are fuller than others. This one is full of air; it's got hollows. If we break it open we shall find little spots where some of the stone is missing* (draws a circle, and another filled with holes). *The grey one is as light as a lump of sugar, because it's a bit empty inside, that's what I think. The other one is more solid, it's full.'*

VOC (11; 8) weighs the two stones and exclaims: *'Oh, I would never have believed it. This one is more solid. One could say the other was hollowed out by the water. This one is hard, there's more matter inside than in the other.*— How so?—*You just have to look. If we break this stone* (the pumice stone), *we'll find lots of holes; it's full of little air holes.*—And the other one?— *There's more stuff inside it than in the other stone.'*

CHAL (12; 0). *'This one is much lighter because it is not so full, not so full of matter.'*

Such are the principal reactions of children at Stage III, who have begun to explain differences in density with the help of both quantitative composition of the weight and the quantity of matter, and have also begun to distinguish between the apparent or global and the corpuscular quantity of matter. Their progress over subjects at Stage IIIA is thus quite obvious, as witness particularly the reaction of Maz. Unlike subjects at the preceding stage, to whom weight is a substantial quality that cannot be quantified and is therefore independent of the volume and the quantity of matter, Maz tried to link the weight to the latter, having first dissociated it from the overall volume by introducing the concept of relative fullness or thickness. However, instead of immediately generalizing this new conception, he still wavered, and, moreover, in a most instructive manner. With the pebble and the piece of wood, he explicitly defined thickness by fullness: the pebble is heavier because it is fuller, and it is fuller because it is thicker. He thus gave the quantity of matter a corpuscular meaning and divorced it from the global volume. By contrast, when he compared the cork to the wood, having first stated that the cork was lighter, he went on to

146

declare that it was nevertheless thicker because it was bigger. In other words, he no longer thought of the quantity of matter in corpuscular terms but once again related it to the global volume. Moreover, this return to earlier notions (which, let it be said in parenthesis, provides us with excellent corroboration of the views advanced in §2) also forced Maz to revert to the treatment of weight as a substantial quality: the piece of wood, although less thick was heavier: 'Sometimes bits of wood are heavier than a cork, because cork is lighter.' Clearly dissatisfied with this solution, Maz then went on to adopt the explicative schema characteristic of Stage III: the wood is smaller and heavier than the cork 'because there are more things in the piece of wood and fewer things in the cork'. In other words, he had reduced the weight to the corpuscular quantity of matter: the wood was heavier because it was fuller. Bae and Debo, on the other hand, provided interesting examples of transitional responses: they still vacillated between the treatment of weight as a substantial quality and as a corpuscular quantity. The stone is heavier, Bae said, because it's 'made of hard stuff', and the Plasticine is lighter, because 'it's got water inside' and hence 'it's not so hard'! Now hardness is still a substantial quality, but it contains a hint about the internal structure: the cork is not hard and therefore not heavy because it 'has lots of holes', and the stone is hard and is therefore heavy because it is 'full of sand'. Debo, for his part, explained that the reason why its hardness made the stone heavier was that 'it's full inside', and that the cork was lighter because it was not so hard, and could therefore 'stretch more easily'. This 'stretching' is comparable to the 'swelling' of the maize invoked by subjects at Stages II and III, but with this difference that 'not hard' has become a synonym for 'lighter' and 'containing less corpuscular matter'.

The concepts of fullness and emptiness introduced at Stage III thus mark the beginning of a differentiation between the corpuscular and the apparent quantity of matter, and also the beginning of the quantification of weight in terms of corpuscular matter. This approach is undoubtedly suggested to the child by his daily experiences with full and empty boxes, and by his examination of various surfaces: a cork has small cracks and a pumice stone looks porous, while the heavier pebble looks smooth and evenly filled with matter. But if this schema is based on daily observations, why does it appear so late? The reason is that the way in which children at Stage III elaborate and employ it is far more complex than the simple model of an empty or full box might suggest: weight has become a quantitative relationship between the number of 'things inside' and the apparent volume. Thus when Hal and Pat said 'there are more things inside and that makes it heavy'; or when Scho explained that 'the stone is heavier because it's full; there's more sand inside' (than the wood or the cork contains elements); or, above all, when Voc said one of the two stones that 'there's more stuff' in one than in the other; they were obviously applying this kind of quantification. The term 'full' thus signifies 'more packed with corpuscular matter' and hence heavier. The inverse of this conception, to which the child sometimes refers by the term 'empty', was used explicitly by Lil, who explained that the pumice stone has 'got hollows. If we break it open we shall find little spots where some of the stone is missing'. To which Voc added: 'If we break this (pumice) stone, we'll find lots of holes.' What we have here once again, therefore, is the 'swelling' schema, but applied to an object of smaller weight and less corpuscular matter than that to which it is being compared.

It is obvious that this correlation of the apparent volume with the number of 'things inside', which marks the beginning of the quantification of density, must sooner or later lead to the schema of compression and decompression: fullness is obviously a pointer to tightness. However, we do not believe that an atomistic approach to density can emerge before the child learns to employ the concept of tightness explicitly. In fact, the atomistic schema of compression and decompression is a schema of separate corpuscles that can crowd into the empty spaces between them or draw further apart, whereas most subjects at Stage III still treat the substance of heavy bodies as more or less continuous, though studded with small holes or cracks (*cf.* Lil). Now while this is undoubtedly a rudimentary form of atomism applied to resistant solids rather than to flour or powder, it is nevertheless of a negative type. Between it and atomism proper we find a whole range of intermediate reactions: thus Scho claimed that the stone was full of sand, and Voc first thought the pumice stone was porous ('we'll find lots of holes') but then went on to say that the pebble had 'more stuff inside' than the pumice stone. But our problem is not to set a rigid limit between successive stages, which would be a completely artificial procedure, but to note both the differences and the similarities in them. In that sense, it is quite evident that Stage III marks a decisive turning point leading up to Stage IV by a series of imperceptible transitions.

§3. *Stage IV : the explanation of density by compression and decompression*

During Stage IV the child not only comes to appreciate that the weight of an object is proportional to its corpuscular quantity of matter, but also that its corpuscles can be packed together more or less tightly, so that at equal corpuscular volume its compression is inversely proportional to its apparent volume. In short, he realizes that its density reflects the fact that its apparent volume varies inversely as its corpuscular quantity of matter :

RIC (10; 8). *'The stone is heavier than the cork.*—Why?—*Because if you put it in water, it will drop to the bottom and the cork won't.*—Why is that?—*Because the stone is tighter.*—What does that mean?—*It's more squashed.*—What is?—*The small things inside.*—How so?—*The stone is made out of squashed sand. The bits of sand are pressed together.*

You say the stone will drop to the bottom of the water. What about the wood?—*It won't.*—Which is heavier, water or the same amount of wood?—*Water.*—Why?—*In the water, it's all pressed tight, but wood has small holes.'*

MART (11; 6). *'The pebble is heavier.*—Why?—*Because of what's inside; a whole lot of little things all squashed together. It's a heap of tiny little stones and small bits squashed together, while the cork is not so tight; it's full of little holes.'*

Ball of clay slightly larger than the pebble:

'Which is heavier?—*The pebble.*—Why?—*It's thicker.*—But isn't it smaller?—*Yes, but if you look carefully, you see they're made differently.*—What are the differences?—*The stone has a little more if you look carefully.*—More what?—*More grains of sand, more bits.'*

Next for comparisons of the pebble with the pumice stone :

MOR (10; 2). The pebble is heavier *'because the bits in it are much heavier than the bits of the other stone.*—Why?—*The sand is all crammed together; it's all in a lump; the sand crystals are stuck together, and that makes it a stone.*—And the other one?—*I think it is also made of bits, but they're lighter.*—Why?—*Because the sand is finer, it crumbles more easily.'*

GIR (11; 2). *'That one* (the pumice stone) *comes from a soft rock.*—What does that mean?—*A rock that dissolves more easily, so it can be split up; the other one is made of a much harder stuff.*—Why?—*It's much tighter, it's made of grains of sand that have been squashed together, like in rocks.'*

BET (12; 0). The pumice stone is lighter because *'the water may have made holes inside* (he taps it). *No, it's not hollow, but it's not so tight.*—And the other one?—*It's heavier because it's tighter.*—Why is one stone tighter than another?—*Because of the air; perhaps the air inside this one* (the pumice stone) *has stopped it from getting so thick* (= tight).—And this one (the pebble)?—*It's much tighter.'* [1]

[1] In connection with these Stage IV responses, we should like to mention the case of a young man who recalled that, when he was a child, he used to explain various phenomena by the same atomistic schema. To him all sorts of things were the result of 'the perpetual wriggling of small grains'. Their relative densities were responsible not only for their different specific weights but also for their consistency, colour, and even sound! Thus he thought that the Adam's apple was a 'small sack of voice grains which escaped from the mouth more or less quickly depending on the force of the breath, each making a little crackle', and that, when a door creaked, 'it let off small grains of iron from the hinge; the faster the door was turned, the higher the pitch and the louder the sound,' etc.

As we see, these children are no longer content with treating density as the relative 'filling' of a given volume by the 'things inside': they define the mode of 'filling', not just by reference to the granular structure of the contents, but also and above all by assuming that the elements composing the total volume are more or less tightly packed together. What we have here is therefore a spatial principle allowing the child to reduce differences in density to the schema of compression and decompression. Thus when Ric was asked to compare the cork to the wood and the stone, he said that the latter was made up of 'more squashed' grains of sand, and even lent this operational system enough coherence to enable him to argue that the elements of water are 'tighter' than the elements of wood, though the former are liquid and the latter solid, and this because wood floats and is therefore lighter than water: 'In the water, it's all pressed tight together.'[1] Mart employed the same schema when he explained that the pebble was slightly smaller than the clay ball because the 'little things' inside are 'all squashed together'. As for the two stones, it might seem at first sight that Mor simply attributed their overall difference in weight to that of their respective parts, since he contended that the 'bits' in one were 'much heavier' than the 'lighter' bits of the other. In fact, however, he, too, was thinking of differences in consistency: the grains of the stone 'are stuck together', 'it's all in a lump', while the 'finer' sand in the pumice stone 'crumbles more easily'. In fact, when it comes to explaining differences in density and no longer, as in the previous chapter, the expansion of one and the same substance, there is nothing irrational in imagining grains of different dimensions. Thus Gir, too, equated density with consistency: the pumice stone 'can be split up', while the pebble was much 'tighter'. Bet, who started out with 'holes' in the manner of children at Stage III, also ended up by discarding relative fullness in favour of compression: 'No, it (the pumice stone) is not hollow but it's not so tight ... perhaps (because of) the air inside.' This transition from the simple notion of fullness to that of a granular structure of greater or lesser concentration shows clearly that the second notion is genetically superior: no inverse development has ever been observed.

But in what precise way is the compression schema the superior of the two? Both schemata seem to involve a spatial composition; both imply invariance and end in a quantification. Why then do the ideas of tightness or looseness never appear before those of fullness or hollowness? While the present series of tests merely reveals the existence of this succession, its nature becomes clear as soon as

[1] *Cf.* J. Piaget, *The Child's Conception of Physical Causality*, Routledge & Kegan Paul, 1930, Chapters VI and VII.

we link the present results to those discussed in previous chapters.

First of all, the belief that the weight of an object is proportional to its global volume and to its apparent quantity of matter is characteristic of all the Stage I responses set out in Chapters 1–7, i.e. of responses based on the denial of all three types of conservation and of all forms of atomism. At that stage, the child invariably relies on perceptive and intuitive appearances, and has not yet learned to apply compositions based on divisions or displacements. When a given solid changes shape, dissolves or expands, the child believes that its substance, weight and volume vary jointly, so that, when he is asked to compare two or three solids of different densities he has no reason to think that their weights may be independent of their volumes, or that their quantities of matter do not depend on their general shape.

Stage II, by contrast, sees the introduction, in all the fields we have been describing, of the quantification and conservation of the substance, while the weight and volume remain intuitive concepts and hence become dissociated from the now constant quantity of matter. Thus when he is asked about the deformation of the clay balls the child will grant the invariance of the quantity of matter, but will argue that every displacement can change the weight and volume of the whole. Again, the grains produced by the dissolution of the sugar are said to be conserved, but to decrease in both weight and volume. During the expansion of the maize seed, the volume is thought to increase but the weight is generally said to decrease: the conservation of matter thus leads to a dissociation of weight; volume and substance and the expansion is considered a simple process of swelling or distension: there is as yet no composition of the volume of the elements. Precisely the same thing happens with objects of different densities: the weight is dissociated from the volume, but remains a qualitative and intuitive concept. As for the substance, it is admittedly composed by means of an extensive as well as an intensive quantification, but since the elements of the substance (no matter whether they are lumps or invisible grains) are not endowed with a constant weight or volume, the child still has no means of distinguishing the corpuscular quantity of matter from the global or apparent quantity. True, when he observes the dissolution or expansion directly, he is forced to adopt some kind of corpuscular hypothesis, and the resulting schemata of fusion or of 'swelling' enable him to reconcile the invariance of the corpuscles with the apparent transformations; but in the case of two bodies of different densities, there is nothing by which he can distinguish the apparent quantities, based on the apparent volume, from the corpuscular quantities; these cannot be deduced from measurements of weight or from compositions of the corpuscular volume, since neither has been quantified. Hence the child limits himself to the assumption that the quantity of matter is by and large proportional to the apparent volume, while the weight remains independent, each substance being endowed with special qualities of heaviness or lightness.

Stage III introduces the quantification of weight and its co-ordination with the composition of the quantity of matter. With the clay ball, the dissolved sugar and the maize seed alike, the child grants the conservation of weight and justifies it by attaching an inherent weight to every particle of the body, so that the conservation of matter automatically entails the conservation of weight. In the case of bodies of different densities this discovery has two essential consequences. In the first place, the child realizes that the reason why the smaller pebble is heavier than the larger cork is that it contains more matter, whence the concept of fullness and the quantification of weight in terms of the number of elements contained in the apparent volume of the whole. In the second place, quantity of matter is differentiated into (1) the more precise concept of corpuscular quantity (which represents the dawn of 'mass'), characterized by the number of elements in, or the fullness of, the object, and associated with the invariant weight of each particle; and into (2) the apparent or macroscopic quantity of matter deter-mined by the apparent volume. However, this mode of composition remains incomplete (and it is this shortcoming which explains the genetic inferiority of the notion of fullness to that of tightness) because the relationship between the weight, or the corpuscular quantity of substance, and the overall volume, or the apparent quantity of matter, has merely been acknowledged but, for lack of a precise composition of the volume, has not yet led to any construction. In fact, the only way of composing the elements with the total volume in which they are contained is to endow every element with an invariant corpuscular volume, and to distinguish, from the sum of the elementary volumes thus defined, a perceptive global volume which varies as the distance between the elements. Now this is precisely what happens in the schema of compression and decompression which appears during Stage IV of the sugar experiment, and which is introduced explicitly at Stage IV of the maize and density experiments.

Our study of the child's conception of density thus shows that after Stage I, during which the child considers the weight of a body as proportional to its global volume and to its perceptive quantity of matter, he goes on to treat the weight as a special quality and dissociates it from the others (Stage II).

During Stage III he once again associates the weight with the quantity of matter, now treated as composed of corpuscles and hence independently of the perceptive volume. At Stage IV, finally, he 'spatializes' this internal quantity of matter with the help of the schema of compression and decompression, which equally implies a corpuscular conception of the volume. In short, while the weight, the quantity of matter and the volume are first fused into a perceptive whole, they become dissociated, only to be recombined into a system of direct ratios, but on a corpuscular plane and in the form of an association between the weight, the mass and the compression, the concept of density thus appearing as the relationship between the internal mass and the apparent volume.

CHAPTER NINE

Special Problems Posed by the Relationship between Weight and Quantity of Matter[1]

The construction of density led the child to a gradual quantification of matter based on the inverse correlation of weight with perceptive volume. In this chapter we shall be looking more closely at the mechanism of that quantification.[2]

In order to avoid a conceptual analysis based on language, we simply ask the child to perform certain actions and to confine his remarks to them. We can, for instance, ask him to make a ball of clay the same weight as a cork, or to build a pile of seeds the same weight as a pile of sand, or finally to make a ball of clay the same weight as half, or quarter the initial cork. Now the importance of these tests is not only that they allow us to check the results of the verbal examinations we have been describing, but also that they reveal the mechanism of the relations used in the construction of density, and quite particularly of their quantification which, as we shall see, introduces a number of curious difficulties.

In §1 we shall examine the child's construction of the inverse relationship between the weight and volume of two bodies of different densities. To that end, we set the child the operational problem of determining changes in volume when the weights remain unchanged. In §2 we shall go on to look at the same problem but in the light of comparisons between one object and a half or a quarter of the other, thus introducing extensive quantifications. In §3 we shall again demand extensive quantifications, but by asking the child to correlate the weights of bars made of different substances with their lengths (at constant breadths).

[1] In collaboration with Trude Strauss.
[2] The reason why we shall be speaking of the relationship between weight and quantity of matter rather than that between weight and volume is that the only volumes the child will be asked to compare in what follows (e.g. a whole cork and half a cork) will be of a global perceptive type, and hence proportional to the apparent quantity of matter. The child is therefore not expected to make any distinction between global and total volumes (Stage IV).

§1. *Three preliminary examples of the correlation of weight and volume*

Our first test is of a purely verbal kind. We show the child two bars, one the colour of lead and the other the colour of iron and ask him: 'Which is heavier, a bar of lead or a bar of iron of the same size?' Most children will reply that the first is the heavier, and those who do not are eliminated from the test. Next we ask: 'In that case, is an iron ball as heavy as a ball of lead? Is it as big, or is it bigger or smaller?'

The second test calls for operational action. The subjects are shown a small, light ball of dry wax and are asked to copy its volume with modelling clay. When they have noted the difference in weight, they are shown a slightly larger ball of wax and are again asked to make a clay ball but this time of equal weight. The original two balls (of equal volume) are left on the table.

What the child is expected to do, therefore, is to proceed from equal volumes (or apparent quantities of matter) to equal weights, and it is this operation (logical multiplication of relations) which we shall discuss in this paragraph. We shall treat the verbal question (first problem) and the operational question (second problem) separately, the better to reveal the differences between abstract reflection and action. Finally, we shall look at a third problem, similar to the second: the construction of a pile of millet of the same weight as a small pile of sand, and *vice versa*.

Now, the responses of our subjects to all three problems proved to be identical. Needless to say, we used one group of children for the wax and clay ball tests, and another for the iron and lead ball tests. The only difference was that the average age at which children progressed from one stage to the next was higher with the verbal questions (iron and lead) than it was with the practical problems (wax and clay or the two piles)—not surprisingly so, since abstract questions are much more difficult to answer. For all that, as we have said, three identical solutions were offered to all three problems, the first two corresponding to Stage I in the construction of density (Chapter 8, §1) and the third to Stage II. In the first two, the child tries to produce two balls or two piles of the same weight by making their volumes equal, or even by making the volume of the heavier larger than that of the lighter! This absurd reaction is produced on the operational no less than on the verbal plane, which shows how difficult children at Stage I find it to dissociate the weight from the volume. The correct solution (volume varies inversely as the density) corresponds to the Stage II responses described in earlier chapters.

Let us first look at a typical Stage I reaction to the verbal test:

PERA (7; 0). 'Which is heavier, a bar of lead or a bar of iron?—*The lead.*—So if a boy wanted to make a ball of iron as heavy as a ball of lead, would he have to make it the same size or not?—*If it's as heavy it will be as big.*—Why?—*So as to have the same weight.*—But which is heavier, iron or lead, when the bars are of the same size?—*Lead.*—So if we made two balls of precisely the same weight how would they be?—*As big as each other.*'

Now for some Stage I responses to the practical problem (wax and clay):

JARO (7; 0). *'They must be as big as each other if they are the same weight.'*
LENK (7; 6). *'They must be the same size, because they are as heavy as each other.'*
JUN (7; 8). *'They'll be the same size; they'll be the same in iron as they are in lead.'*
KIR (8; 9). *'They'll have to be made the same size.'*
RIC (4; 11). 'Take this (wax) ball and weigh it in your hand.—*Gosh, it's light!*—Yes, it's made of wax, it's very light. Now make me a ball the same size with this clay.' He does so. *'Here you are.*—Well, now weigh it in your hand. Which is heavier?—*This one* (the clay). *It's heavier because it's made of clay.*—Now, listen carefully. Here is another ball of wax. Now make me a clay ball that will be just as heavy – listen carefully – just as heavy and no heavier than this ball of wax.' Ric gathers up all the modelling clay he can find to make a ball of the same size as the clay one and says: *'It'll take this bit as well.'* He scrutinizes the result and adds some more. 'What are you doing?—*I'm trying to see if it's as big.*—Should it be as big?—*Yes.*—Why? —*To get it as heavy.*—Which of these two (the two balls from the first experiment) is heavier?—*That one* (clay).—So if these (the two he is working with) are as big as each other will they also be as heavy as each other?—*Yes.*—Why?—*Because they're as big as each other.'*
AR (5; 0) is shown a ball of wax that is bigger than a ball of clay but of equal weight. He says that the bigger one will be heavier: 'Make sure.' He weighs them. *'No, they're the same weight.*—Why?—*One of them is hard and the other one isn't.*—Which one is hard?—*The small one* (which is soft clay, but "hard" clearly means heavy or compact).—Well then, here is a small ball of wax. Make another that weighs just as much out of the clay.' He gives it the same volume. 'Will it weigh as much?—*Yes.*—Why?—*It's just like that one* (the small ball of wax); *it's just as small.'*
GAS (6; 0) is asked to make a ball of clay the same volume as the wax ball, and does so: 'Which is heavier?—*That one, the clay ball.*—Why?—*If it's as big, it must be heavier.*—Very well, now look. Here is another ball of wax (a little larger). Make me a ball of clay that weighs as much.' He makes it the same volume. *'Here you are.*—Is it the same weight?—*Yes.*—Why?—*They're as big as each other.*—But you've just told me that when they're as big as each other the clay one is heavier.—*Yes.*—So?—*Yes, I remember* (embarrassed expression).—What must you do?—*I don't know.'*

And now for the two piles (sand and millet):

NIER (5; 4). 'Look, here is a small box of sand and a small box of seed

156

(the two boxes are of the same size). Which is heavier?—He weighs them. *'The sand is heavier, and the seed isn't so heavy.*—Well then. Take some sand from here and some seed from over there. Now make me a pile of sand as heavy as that pile of seeds, and a pile of seeds as heavy as that pile of sand.' He constructs two piles of the same size. 'Which of these two boxes was heavier (the control boxes left on the table)?—*The one with the sand.*—And are the piles you have made as heavy as each other?' He examines the piles to see if they are exactly the same size. *'The seeds are heavier, the pile is a little bigger, I mean this pile of seed* (he adds some sand to the pile).—And now?—*Now, they're as heavy as each other.'*

NAD (5; 0) constructs two piles after weighing the two boxes: 'Will they have to be the same size, or will one have to be smaller?—*The same size.'* He finishes his construction. 'Are they as heavy as each other?—*Yes.'* He weighs them and discovers that he is wrong. *'Oh no, the sand is heavier.*— So?—*We need a little more of the seeds.'*

CAR (6; 0) ascertains that the sand is heavier but constructs two identical piles: 'Are they as big as each other?—*Yes.*—Why did you do that?—*To make them as heavy as each other.*—But when you weighed the two boxes were they the same?—*The sand was heavier.*—So in order to make two piles as heavy as each other don't we have to put in more sand or more seeds?— *No.'*

TRE (8; 0) ascertains that the sand is heavier but nevertheless makes two piles of the same volume: 'Why have you made them the same size?—*We want them the same weight; so I've put in the same amount of seed.*—But did these (the two boxes) hold the same amount?—*Yes.*—And were they the same weight?—*No.*—So why have you put the same amount here?—*So as to make them as heavy.'*

These responses are open to two interpretations. Either the child forgets what he has just ascertained for himself (and Gas's reaction suggests that he did just that), or else he does recall the data but fails to invert the relations: 'A is heavier than B at equal volumes' into: 'A is less voluminous than B at equal weights'. Now, in small children the failure to make this inversion can easily entail a loss of data, but this would not, in itself, validate the first interpretation, because our problem is to decide whether or not forgetting the data is the result of intellectual difficulties.

Now, while the issue is uncertain when we question no more than one or two children, the great frequency of this type of reaction and, in its absence, the appearance of the second type of response we have mentioned, suggests strongly that, at Stage I, forgetfulness and distraction cannot be the full explanation, and that the logical difficulty impeding the multiplication of the relations must be the essential factor.

Let us note, first of all, that all the responses we have just cited involve the same reasoning. The subject begins by ascertaining that, at equal volume, the weight of A is greater than that of B. If A $\stackrel{\scriptscriptstyle =}{\scriptscriptstyle \cdot}$ B expresses equality of volume, and A $\xleftarrow{w}$ B that the weight of A is greater than the weight of B, we have $(A \xeq{\underline{L}} B) = (A \xleftarrow{w} B)$. Now, from this system the child concludes, when asked to predict the weight of the iron and lead balls or to construct a clay ball equal in weight to a ball of wax, or a pile of sand equal in weight to a pile of seeds, that $(A \xeq{\underline{L}} B) = (A \xeq{\underline{w}} B)$, thus denying his original premise. When he is asked to predict the respective weights of the balls of lead and iron in the abstract, this error persists until a later age than it does on the practical plane, but in either case it reflects the same logical structure. Why should this be the case?

We might say that our subjects fail to see any connection between their initial discovery that the weights are unequal, i.e. that $(A \xeq{\underline{L}} B) = (A \xleftarrow{w} B)$, and the subsequent problem of quantification, i.e. the construction of two piles or two balls of equal weight. In the first case, having discovered that at equal volumes the weights are unequal, the child concludes that weight is a substantial quality more characteristic of the lead, clay or sand than of the iron, wax or millet, and hence fails to pose the problem of the quantification of a quality he has defined in a purely egocentric and phenomenalistic manner (see Chapters 1–6). In the second case, by contrast, he is asked to solve the quantification problem from without, i.e. to construct a ball or a pile (A) of the same weight as the ball or pile (B) and to determine its volume. But unable to quantify the weight either as such or in relation to the volume, he simply tells himself that to obtain equality of weights he must equate everything else as well, i.e. he poses $(A \xeq{\underline{w}} B) = (A \xeq{\underline{L}} B)$.

But why does the quantification of weight and volume elude children at this stage of development? This is our real problem, and there are two possible answers. We can either invoke physical reasons and say that, at this stage, weight is no more than an intuitive, and hence egocentric and phenomenalistic quality, which is thought to change with changes in shape and hence does not lend itself to measurement and quantification; or else we can invoke logical reasons: every quantification demands a logical composition, and the child at this stage cannot yet group weight relationships, let alone multiply them. But which of these two explanations is the correct one? Is it his lack of physical knowledge which prevents the child from constructing a logic of weight, or is it his lack of logic which ensures that his conception of weight remains so intuitive and so unphysical?

It is quite possible that these two interpretations are but one. The physical operations of division and displacement are nothing but logical operations in time and space, and applied successively to various physical qualities: to the substance first of all, and then to the weight, volume, etc. There is no doubt at all that this happens with the substance, or more precisely with the apparent quantity of matter, a concept whose construction coincides with that of quantity in general, and which is thus the first form of quantification. Now, at Stage I the substance is not yet quantified, let alone the weight; it is only during Stage II that the quantity of matter is first treated as an invariant, and that the weight can be dissociated from it. The quantification of the latter poses a new problem, the solution of which is not discovered until Stage III. We may therefore take it that, at the level under review, quantification is quite impossible for physical no less than for logical reasons.

Now, in addition to the responses we have given, there is in fact another type of reaction: when the child discovers, against all his expectations, that the weight of A is greater than the weight of B, he concludes that in order to construct a pile or a ball of A equal in weight to B he must make A larger than B! Here, first of all, are a few responses to the verbal problem (lead and the iron):

RET (6; 0). *'The lead is heavier than the iron.'*—Well, then, if we wanted to make a ball of lead as heavy as the iron one, would we have to make it the same size or what?—*We'll have to make one of them bigger.*—Which one?—*The lead, because lead is heavier than iron.'*

HAL (7; 6). 'Is a bar of iron heavier or lighter than a bar of lead?—*Lighter.*—So if we wanted to make a ball of iron and a ball of lead both as heavy as each other, how big would they have to be?—*We must make the lead bar bigger, because it's heavier.'*

NOY (8; 0). *'The iron will be smaller because it's lighter.'*—So will one be heavier than the other?—*No, they'll be the same if we make the lead bigger.'*

Now for the wax and clay balls:

BAD (5; 8). 'Make me a ball as big as that one.' He does so. 'Which is heavier?' He weighs them. *'That one, because it's made of modelling clay.*—Well, then, here is a large ball of wax. Make me a ball of clay precisely the same weight.—*It must be bigger.'* He makes a ball of clay of exactly the same volume as the ball of wax. 'And will this one be as heavy as the other?—*No, it would have to be bigger.*—Why?—*To have the same weight.'* He adds some more clay and makes a much larger ball. 'And now?—*It'll be the same weight.'* He weighs it. *'Oh, no, it's heavier!'*

BOUL (6; 0). 'Make me a ball the same size.' He does so. 'Which one is heavier?—*The clay ball.*—Now look at this ball of wax. Make me a clay ball of the same weight.' He gathers up all the lumps of clay on the table. 'What are you trying to do?' He finishes up with a ball larger than the wax model. *'I must make it as heavy as that one.*—And do you think you have?—*Yes, but I'm not sure.*—Why not?—*I think there isn't enough clay. It ought to be bigger to weigh the same.'* He is handed some more clay and he makes an even larger ball. 'Weigh it.—*Oh, it's heavier!*—Why is that?—*Because there is more clay.*—What can you do about it?—*Use the same amount of clay.'*

And with the two piles of sand and millet:

KAD (6; 0) states that the sand is heavier than the seeds. A small pile of sand is poured on to a tray and he is asked to make a pile of millet of the same weight. He does not use enough seeds: 'What are you trying to do?—*We need more sand because sand is heavier.*'

MAG (6; 0). Same reactions: 'Weigh it.—*Ah, I have used too much sand.*— Well, try to put it right.' He removes some sand, but not enough. 'Which of the two piles is bigger now?—*It's still the sand.*—Why?—*To make it as heavy.*—But which one was heavier, the box of sand or the box of seeds?— *The sand.*—So why do you make the pile of sand bigger?—*Because sand is heavier.*'

JUC (7; 6) first makes two piles of the same volume. 'Why did you do that?— *They have to be the same size to be the same weight.*—But which one was heavier in the box?—*The sand.*—So what must you do to make the two piles the same weight?—*Ah, I must put in some more sand.*' He does so. 'What are they now?—*As heavy as each other.*'

It is quite obvious that forgetfulness cannot explain the responses of these subjects. True, with the iron and lead balls, the verbal nature of the test may have been responsible for a purely mechanical association of one 'more' with another. But Bad, Boul, Mag and Juc constructed their argument step by step and made obvious efforts to dissociate the weight from the volume, once observation had shown them that the two were independent of each other. Hence it could only have been their very attempt to dissociate the weight from the quantity of matter that forced them to revert, as it were despite themselves, to the idea of direct proportionality. Bad, for example, began with a clay ball the same size as the wax, but corrected himself and said that 'it must be bigger ... to have the same weight', which was plainly not the result of a verbal association of one 'more' with another but of a vain effort at co-ordination. Boul even thought that his ball was not big enough and added some more clay, saying: 'It ought to be bigger to weigh the same.' In other words he assumed that, at equal volumes, the two weights would differ, and concluded that a larger quantity of the heavier of the two substances must be used to make it the same weight as the lighter! Mag was surprised to find on the scales that he had put in too much sand, but nevertheless kept much too large a pile of sand so as 'to make it as heavy ... because sand is heavier'. Juc, who began with two equal piles, suddenly recalled that the sand was heavier and concluded that 'I must put in some more sand' to equalize the weights. In a sense, therefore, it is true to say that the 'more' entails the 'more' for these subjects, though not by verbal association but because of their inability to multiply the relations and above all to grasp the inverse relationship between weight

160

and volume. Their argument is, in fact, so 'unverbal' that they actually construct a larger ball of clay or a larger pile of sand to reproduce the weight of the ball of wax or the pile of millet, while saying that, at equal volumes, the first substance is heavier than the second!

How are we to explain these extraordinary reactions? Let us note first of all that there is nothing exceptional about them. In dealing with time and speed, two concepts to which we hope to devote a special study, we also discovered that children think that a fast-moving body will take more time to cover the same distance as a slow-moving body, or that a body traversing the smaller of two concentric paths travels more quickly than one covering the larger path in the same time. Hence there is nothing peculiar about the two types of response we have just mentioned. In particular both involve two parallel 'explanations, one physical and the other logical. Let us look at the physical interpretation first. If we are right to think that children at this stage of development are incapable of quantifying weight, volume and even the apparent quantity of matter and hence of grasping their conservation, i.e. that they fuse all three into an egocentric and phenomenalistic whole, then they are bound to conclude from the fact that a substance (A) is heavier than a substance (B) at equal volumes either that (A) has a substantial quality that has no bearing on the ratio of the weight to the volume (the first reaction); or else that (A) is quite generally endowed with 'more' of everything. The reader may recall the responses of Zur and Che (Chapter 8, §1) who predicted that a large cork would be heavier than a small stone, and who, once they were undeceived, went on to say that the stone was bigger than the cork (Zur); or that a silver franc was heavier than a brass franc because it is thicker when it had only just been said to be thinner (Che). Similarly, our present subjects undoubtedly believe that the heavier substance A must have something 'more' than the lighter substance B, the 'more' being something undifferentiated and global. Hence, when they are asked to construct a ball or a pile of A having the same weight as B, they think that because A has more weight than B, it must also be more voluminous. Thus when Kad said, 'We need more sand because sand is heavier', he was merely expressing his failure to distinguish between the quantity of matter and the weight; and it is this very failure which explains his and similar bizarre constructions: it prevents the inversion of the two terms. In short, the child (1) anticipates that the weight will be proportional to the quantity of matter, i.e. that $(A \stackrel{L}{=} B) = (A \stackrel{w}{=} B)$, but discovering (2) that the weight of A is greater, i.e. that $(A \stackrel{L}{=} B) = (A \stackrel{w}{\leftarrow} B)$, he decides that, in order to balance the weights, he must (3) increase the quantity of matter of A, i.e. construct $(A \xrightarrow{w} B) = (A \xleftarrow{v} B)$.

It is obvious that, side by side with this physical explanation, we must also look for a logical explanation. This poses no fresh problem : the failure to conclude from $(A \xrightarrow{v} B) = (A \xleftarrow{w} B)$ as to the inverse relation $(A \xrightarrow{w} B) = (A \xrightarrow{v} B)$ is simply a particularly striking example of the difficulty with which reversible compositions face subjects at this stage. Now, even without performing such operations (groupings of bi-uniform multiplications of asymmetric relations)[1] the child could easily arrive at an intuitive or empirical solution. The fact that he fails to do so shows plainly that he has not yet learned to replace the system of egocentric relations with one of reversible operations.

The links between the physical and the logical factors thus seem very close once again. Needless to say, we must distinguish the practical reactions (wax and clay or sand and millet) in which the interaction of physical representations with reason is quite obvious from the purely verbal reactions (iron and lead) which are residual and persist much longer. We must also remember that once the quantification of substance has been constructed, the logical mechanism is ready for use, though it cannot be extended directly to weight and volume which raise new physical problems.

All in all, the two types of Stage I reaction thus constitute a homogeneous whole, whose common principle – the non-differentiation of weight, volume and substance – manifests itself logically as an inability to proceed to reversible compositions and quantifications. Stage II, by contrast, sees the discovery of the correct solution thanks to the dissociation of the weight from the quantity of matter, the second having become open to composition while the first remains intuitive. As we saw in the last two chapters, this dissociation brings home to the child the inverse relationship between weight and volume, and hence enables him to solve the three problems with which we have been concerned in this paragraph. But, as we shall discover in §2, this inverse relationship can also be discovered intuitively, for at Stage II it does not yet imply a quantification of the parts (half or quarter) of an object. Before looking at this point more closely, let us first set out a few typical Stage II responses, beginning with comparisons of the lead and iron balls :

DON (6; 6). *'The iron ball will have to be bigger if it is to be as heavy as the lead.'*

HUM (7; 6). *'We must make them the same size. No, the iron ball must be bigger, because we need more iron to make them as heavy as each other.'*

[1] See *Proceedings of the Société de Physique et d'Histoire naturelle de Genève*, 1941, 58, p. 154.

ONS (8; 0). *'The lead ball must be smaller, because it's heavier.'*

RAD (9; 0). *'The iron ball is lighter than the lead, so we need more of it to make it as heavy.'*

GAR (10; 0). *'The iron ball will be bigger, because we need more of it to make the same weight.'*

Now for the practical problem (wax and clay):

MOR (6; 0). 'Make a ball the same size as this wax ball here.' He does so. 'Which one is heavier?—*The clay ball.*—Why?—*I don't know.*—Look at the first wax ball. What would you have to do to make precisely the same weight?—*I'd have to make it smaller because this one* (the model) *is made of wax.'*

COR (6; 6). *'We'd have to make it smaller.*—And what if we made it as big?—*It would be heavier.'*

ANO (6; 6). *'We'd have to make it smaller.*—Why?—*Clay is heavy, so we need less of it.'*

MEY (7; 0). *'We need less modelling clay, because it's heavier.'*

And with the two piles:

TON (6; 6). *'The sand is heavier, but it's as big.*—Well, then, can you make two piles as heavy as each other, one with the sand and the other with the seeds?— *I'll have to make the sand pile smaller because it's thicker* (= heavier). —Do it then.—*I need a bigger* (= more voluminous) *pile of grains because they're not so heavy.'*

REL (6; 0). *'The sand is heavier, so the pile of grains ought to be bigger, because the grains are not so heavy.'*

MAZ (6; 6). *'We need more grain.*—Why?—*Because the sand is heavier.'*

MAY (7; 0). *'The sand is heavier, so we need to make the pile of seeds bigger.*— Why?—*So that they both weigh the same.'*

All these responses reflect the correct dissociation of weight from the quantity of matter in purely physical terms, which agrees with the general characteristics of Stage II. They also reflect a logical capacity for co-ordinating two inverse relations. However, since the substance is already quantified at this stage, both intensively and extensively, its inverse correlation with the weight does not demand a new operation and may, in fact, be the result of purely empirical or intuitive progress. In other words, the weight as such remains an intuitive quality, not subject to special quantification. Now this is, indeed, what we shall discover in §2: at Stage II the inverse relationship between weight and volume is only applied to whole objects, but not to half or quarter objects – clear proof that though matter as such can be quantified extensively, weight has remained a substantial quality. That is precisely why we have described all the reactions we have just discussed as Stage II responses, and why more complicated tests are needed to distinguish the latter from Stage III responses.

163

§2. *Cork and clay*

The child is shown a large cork, broader than it is tall (the kind of bung that used to be sold for pickling jars), and various loose bits of modelling clay. He is asked to turn the clay into 'something as heavy as the cork', and once he has done so he is asked to check his answer and to make what corrections may be needed. Next he is shown another cork of the same dimensions but cut vertically into two and he is asked to make a clay figure of the same weight as one of the two halves. Finally, he is asked to do the same for a quarter of the original cork.

He thus has to solve a double problem: not only must he dissociate the weight from the volume, i.e. remember the density of the cork and of the clay (*cf.* §1), but he must also be able to break down the apparent weights into halves and quarters (quantitative decomposition).

Now from their responses we shall see that children at Stage I do not dissociate the weight from the volume: their clay constructions tend to be of the same shape and dimensions as the half and quarter corks. Next comes an intermediate stage (Stage IIA) during which they predict that a ball of the same dimension as the cork will be heavier, but do not base their constructions on these predictions. At Stage IIB, they correctly dissociate the weight of the undivided cork from its volume and produce a clay ball that is smaller than the model, but when it comes to copying the weight of the half and quarter corks they ignore the quantitative relations: instead of simply dividing their own constructions into two or into four, they produce new balls of arbitrary volume. Finally, at Stage III they produce the correct quantifications (this test does not call for Stage IV responses).

We shall begin with several Stage I responses:

RIC (4; 11) makes a clay ball almost the same size as the cork: 'Are they the same weight?—*No.*—Which one is heavier?—*The cork.*—Why?—*Because it's a little bit bigger.*—Weigh it.' He does so. '*Oh, it's too heavy.*—So what must you do?—*I'll have to take some off.*' He removes a piece. '*No. I've taken off too much.*' He puts back exactly what he has just removed.

After a series of corrections, he produces the same reactions with the half and quarter corks.

EB (5; 7). 'Is the cork heavy?—*No.*—And the clay?—*That's heavy.*—Well, make me a piece of clay just as heavy as this cork, no heavier, and no lighter.' He sets to work. 'To be as heavy, will it have to be as big, or won't it?—*It's got to be the same size.*—Why?—*Because the clay is heavier.*—So, if it's heavier, what must you do?—*I must make it as big.*' He does so. 'And will it have the same weight like that?—*Yes.*—Why?—*Because it's as big.*' On weighing he discovers that the clay is heavier, and repeats his mistake

164

when trying to reproduce the weight of the half cork.

REN (5; 7). 'What will we have to do to get a ball of the same weight?—*Make it as big as that one* (the cork).—Is clay as heavy as cork?—*No, clay is heavier.*—Why?—*Because cork is lighter.*—Well, then, make me a ball of clay just as heavy as the cork.' He makes it the same size. *'There you are.'* He weighs it and discovers that the clay is heavier. *'No, it's wrong'.* He picks up a fresh bit of clay and makes it the same size as the first. *'Now it's right.'* He weighs it and discovers his mistake. *'Oh, dear!*—What must you do now?—*I must take some off.'* He makes a big hole in the clay but removes nothing and weighs it again. *'It's still too heavy, I'll have to make it smaller.'* He removes a piece.

'And now make me something as heavy as that one (the half cork).' He adds some clay and copies the shape and size of the model. 'Weigh it.' He corrects his construction step by step. 'And what about this one (the quarter cork)?' Same reactions.

AND (5; 10) takes his time looking at the cork, and then picks up several bits of clay with the obvious intention of producing a figure of the same volume. The end result is a smaller ball, of which he says: *'It'll weigh less.*—Why?—*It's not so high.'* He adds some more clay and ends up with a figure more or less the same size as the cork. He weighs it and finds that it is too heavy; 'What must you do now?—*Take some off.'* He removes a tiny fraction, re-weighs the remainder, and so on several times. Same reaction with the half and quarter corks.

LUG (6; 6). 'Take some of this clay and turn it into something that will have the same weight as that cork.' He makes a disc with the same diameter as the cork, and applies it against the model to check. Then he adds a second disc, again measuring the diameter of the cork. 'Why do you make these discs?—*To make it as round as the cork.*—Do you think they will weigh the same?—*The cork will be heavier because it's bigger.*—Well, make them the same weight.' He adds another disc and says: *'I'll make it just like that one* (and points to the cork) *then it will be the same weight.*—But why do you take one disc after another?—*Because I can't get it right straight off.'* He ends up with a copy of the cork, weighs it and looks astonished to discover that it is heavier. He removes disc after disc, weighing the results with mounting astonishment, and finally says: *'Gosh, that modelling clay is heavy.'*

Having at last discovered the equivalent weight, he is shown the half cork, but instead of cutting his own construction in two, he once again adds new pieces of clay to copy the volume of the model. He weighs the result and then exclaims: *'Oh, no. It needs less.*—Why?—*Because half the cork doesn't weigh such a lot.'*

All these reactions, like those described in §1, reflect a total lack of differentiation of the weight from the volume or the quantity of matter. Thus when Ric ended up with a ball of clay that was barely smaller than the cork model and thought that the cork would be heavier 'because it's a little bit bigger'; when Eb and Ren, after stating that clay is heavier than cork, nevertheless claimed that they would have to construct something 'the same size'; when And thought that his ball would weigh less than the model because 'it's not so high'; and when Lug painstakingly piled up discs of clay, each checked against the model, it is clear that all these children had not yet begun to dissociate the weight from the volume and the quantity of matter, thus confirming the conclusions we set out in §1 of the last chapter. Now this confirmation is the more important in that the reactions described in Chapter 8 might have been mistaken for purely verbal responses: we now see that children at Stage I apply the same principles to their actions.

What is so remarkable about these new data is that they adduce clear proof that the direct proportions which children at this stage establish between the weight, the volume and the quantity of matter in no way reflect a need for precise quantification but are, on the contrary, based on a subjective intuition of perceptive equivalences. The responses of Eb and Ren are particularly telling in this respect: both had stated explicitly that the clay was heavier than the cork, but nevertheless concluded that, to obtain the same weight the piece of clay would have to be 'the same size', precisely because, as Eb explained 'the clay is heavier'.

We have already tried to interpret these absurd reactions in physical as well as in logical terms (see §1), and we must now complete our analysis by showing why neither the weight nor the volume can be quantified at this stage. When the child says that the clay is heavy and the cork light, he is only thinking of his subjective impressions (the pressure the two substances exert on his hand) and never even suspects that these impressions could be measured or evaluated by objective means. Moreover, when he is asked to make something of the same weight, he simply tries to produce as perfect a copy of the model as he can, because he believes that perceptive equivalence must go hand in hand with weight equivalence. Eb's argument: 'it's got to be the same size because the clay is heavier' thus ceases to be absurd and simply means that, since the clay and the cork are different qualities, the model must be copied as faithfully as possible if the result is to produce the same impressions. Similarly when Ren, realizing that the weights were unequal, said: 'we must take some off', he was not so much thinking of changing the quantity of matter as of changing the shape: hence the hole he made in the clay.

Nor is that all. When these children, having discovered that the piece of clay must be much smaller than the cork, go on to copy the weight of half or quarter the cork, they completely forget their own discovery and instead of simply cutting their balls or lumps of clay in two, once again try to copy the volume of the model. Lug even went so far as to add more clay to the small piece whose weight he had just found to be equal to that of the whole cork, thus acting as if the half were heavier than the whole! This systematic mistake, which continues throughout Stage II, suffices to demonstrate the non-quantitative nature of the child's conception of weight and volume in the early construction of density.

During an intermediate sub-stage (Stage IIA), to which we must now devote a few words, the child does succeed, as soon as he discovers that the clay is heavier than the cork, in co-ordinating the weight with the volume or with the quantity of matter, and concludes that the lump of clay must be smaller than the cork if it is to have the same weight. But when he tries to put this idea into practice, he sooner or later reverts to non-differentiation and again merely copies the volume of the model. Here are a few examples:

SUM (5; 11). The cork *'is not heavy.—And the clay?—It's heavy.—*Will you make me a piece of clay that's as heavy as the cork?—*Yes.—*Will it be as big?—*No, it won't.—*Will it be bigger or smaller?' He picks up the clay and produces a copy the same size as the cork. *'The same.—*Will it be the same weight?—*No, this one* (the clay) *will be heavier.—*What did I ask you to do?— *To make them the same weight.—*Well, why don't you?' He picks up a fresh piece of clay and copies the volume once again. 'Will that be right?—*The cork will be lighter.—*Can you put it right?' He keeps removing clay and eventually obtains the correct solution. 'And what about this one (the half cork)?' He adds some clay and copies the shape and volume of the model.

DIN (6; 11) says straightaway that *'the clay is heavier.—*So, to make a lump the same weight as this cork, must we make it the same size, or what?— *We must make it a little smaller.'* But he nevertheless produces a lump the same size. 'Will it weigh more or less than this cork?—*The same* (weighs it). *Oh no. It's heavier.'* He keeps removing clay until the two are equal.

'What about this one (half cork)?' He copies the exact volume. 'And that one (the other half)?' Same reaction. 'What if we weighed this lot (the two lumps of clay representing the volume of the two half corks) against that one (the whole cork)?—*The clay will be heavier because it's in two lots and that one* (the cork) *is in one.—*And if we weigh this lot (i.e the clay cut into eight pieces)?—*The cork will be heavier, because these little bits come to nothing.'*

KEL (7; 0) says: *'We must make the clay smaller',* but as soon as he has finished his small lump, adds more clay until he obtains a figure the same volume as the cork. He then weighs the result and says: *'It ought to be smaller.'*

With the half and quarter corks, he copies the volume straightaway. After repeated weighings, he obtains a lump the same weight as the quarter cork. The demonstrator now places two quarter corks on one pan of the balance and hands the subject the lump he has just made, and another of the same weight: 'Make a lump the same weight as the two corks together.' Instead of using the two lumps he has been offered, he takes just one and adds a large quantity of clay at random.

These Stage IIA reactions are of great interest, because they show that subjects at this level of development are so caught up in the schema of non-differentiation that even when they themselves declare that the ball of clay must be smaller than the cork if it is to be of equal weight, they go on to produce a lump the same shape and dimensions as the model. In other words, the tyranny of the perceptive schema still carries the day over the co-ordination of relations. Their composition of the parts into the whole thus marks no progress over Stage I: to reproduce the weight of half the cork, they continue to add clay to the original lump, equal to the weight of the whole, and they do the same for the quarter, despite everything they have learned from the first weighing. So ignorant are they of the laws of composition that they believe eight-eighths weigh less than two halves, or that the whole cork weighs more than eight-eighths (Din), or that to make a piece of clay equal in weight to two quarter corks they must add an arbitrary quantity of clay to the piece representing the weight of one of the quarters (Kel).

In short, these subjects, like those at Stage I, do not even suspect that the weight bears a quantifiable relationship to the volume or the quantity of matter; they simply treat it as a subjective quality, such as a colour or a smell, and though they start out with the belief that the weight must be dissociated from the apparent quantity of matter (the volume), they fail to express this idea in their practical operations. They thus herald the next stage, even while adhering to the first in practice.

Children at Sub-stage IIB, by contrast, give practical expression to their belief in dissociation: not only do they predict that the clay will be denser than the cork, but when asked to make a lump of clay the same weight as the cork, they straightaway try to construct a smaller lump. Like the children at Stage II described in §1 of this chapter or in §1 of Chapter 8, they therefore assume that an object can be both smaller and heavier than another, but still fail to quantify the weight they have dissociated from the quantity of matter and the volume, and continue to treat it as a simple quality. Thus, when they are asked to construct balls of clay corresponding in weight to a half or quarter cork, they do not simply divide their original constructions into two or four, nor do they make new lumps with these proportions, but construct lumps of arbitrary size, sometimes larger than the original lump, and in any case much too big:

ALE (6; 0). 'You see this piece of clay? Turn it into something that weighs exactly the same as this cork.' He picks up a lump as large as the cork and immediately removes a part. 'Why did you do that?—*Because the clay is harder and heavier, it lifts up the cork'* (on the balance). He finishes with a very small lump of clay and says: *'Now it's as heavy.*

'And what if I cut this cork into two?' (He looks at the half cork and compares it with the small ball equivalent to the weight of the whole. *'We must add (!) a little.*—Why?' He weighs it. *'No, it's too heavy.'* He keeps removing clay until he obtains the correct weight. 'And what about a quarter of the cork?—*We must take some off because it weighs less.'* But he does not simply cut his own lump in two.

BAN (6; 0) immediately produces a ball A smaller than the cork, then reduces it further to B, and finally to C (correct). 'And now make me a ball the same weight as half the cork.' A half cork is placed before him. Instead of cutting C in half, he picks up a fresh piece of clay, and looking at the half cork, constructs a ball a little smaller than the half cork, but almost the same size as A. He weighs it, and removes some clay, but leaves the new ball much larger than C.

After further corrections he is asked to copy the weight of the quarter cork. He says: *'The ball will be smaller,'* but simply removes a tiny piece without bothering to weigh the remainder.

POR (6; 10) makes a ball appreciably smaller than the cork, then, finding it still too heavy, squashes it in an obvious effort to reduce the weight (thus clearly showing that he does not believe in the conservation of weight, as, indeed, no subjects at Stage II do). When he has removed a succession of fragments to obtain the required weight, he is shown a cork cut into two halves: 'What kind of ball would weigh the same as these two halves?—*It'll have to be bigger than before because there are more pieces now.'* He makes a larger ball and weighs it, then reduces it and finally says: *'Oh, no, it's the same.*—So the whole cork and these two halves weigh the same as this ball. Now if I take one of the halves away and ask you to make a ball as heavy as the other, what would you do?—*I'd make it smaller.*—Smaller by how much?—...—A lot smaller?—*I don't think so.*—Then how much smaller?' He removes a little of the clay, and keeps doing so until the scales balance: 'Here you are.—And with this quarter cork?—*We must make it even smaller.'* But he removes far too little.

KES (7; 7) picks up a smaller lump of clay. 'Why?—*Because it's less heavy, so I've taken a bit less.'* He keeps removing clay until he obtains the right weight. 'And what about this one (the half cork)?' He picks up a lump of clay corresponding to the weight of the entire cork, weighs it in his hand and then says: *'It'll be too much.*—What must you do?' He cuts off about one tenth. *'There you are.*—And with this one (the whole cork)?' He puts back the piece he has removed. 'What have you done?—*First, I cut off a little bit, and now I've put it back.'* The same response with the quarter cork.

Next he is shown the whole cork and a lump of clay of the same weight, together with an identical cork cut in half. 'What must we put down for this one (first half)?' He removes about one-eighth of the ball corresponding to the whole cork, and points to the remainder (i.e. seven-eighths). *'This.*—And for that one (the other half)?' He reduces the one-eighth he has just removed. 'Do you think that's right?' He removes five-eighths from the original seven-eighths. *'Like that.*—And for this one (he is given back the whole cork)?' He combines the two small balls, i.e. the one-eighth and the two-eighths, neglecting the residue. 'And what does this (the residue) go with?—*I don't know.'*

169

TAL (9; 0) similarly has no difficulty in producing a small ball corresponding to the weight of the whole cork, but when asked to copy the weight of the half cork, he first removes part of his ball and then, for the other half cork, copies his construction. When asked to construct the weight of the two half corks together, he combines his two balls which, together, weigh roughly one-and-a-half times as much as the whole cork.

POLI (9; 2) produces a *'smaller'* ball for the whole cork: 'Why?—*The clay is heavier.*—And what about this one (the half cork)?—*I must take a bit away.*—How much?' Removes a piece at random. 'And if we take these two bits (the piece he has just removed and the remainder) will that give the same weight as that (the whole cork)?—*No. One of the two is smaller.*—So?—*It won't be as heavy.'* For the quarter cork, he removes another piece, but he is incapable of reconstructing the whole by combining the parts.

These responses are most illuminating for, when considered in conjunction with those produced at Sub-stage IIA and at Stage I they cast fresh light on the non-conservation of weight and volume. Throughout this book, we saw, both in connection with the conservation of weight (Chapters 2, 5 and 6) and also in connection with the role of weight in the explication of density (Chapters 7 and 8), that the quantification of weight presupposes a reversible composition such that a given whole can always be broken down into fragments that conserve their weight, no matter what their arrangements or their displacements. Now this construction, precisely because it forces the mind to treat conservation as a deductive necessity, presupposes that the child has already acquired a grasp of the concept of quantity of matter, though obviously not yet of its relationship with weight. When it comes to the quantification of matter, in fact, the child at Stage II knows perfectly well that the sum of the parts is equal to the whole, that two halves or four quarters contain neither more nor less substance than the whole: it is precisely the construction of this total invariance which makes him realize that a ball of clay must be smaller than a cork of equal weight. However, while he has learned to apply the qualitative groupings of classes and relations and the additive and multiplicative groups of numbers to the composition and quantification of the substance, he is not yet capable of applying them to the weight. In Chapter 2 we saw that children at Stage II do not realize that a coil of clay conserves its weight when it is cut in two. We now see, but much more clearly, that the child at this stage is quite incapable of additive compositions in respect of the weight and the quantity of matter. Thus if Q is the quantity of clay corresponding to the weight W of the whole cork, the child never concludes that $\frac{1}{2}$ Q is equal to $\frac{1}{2}$ W, or that a $\frac{1}{4}$ Q is equal to $\frac{1}{4}$ W! Though he knows perfectly well that $\frac{1}{2}Q + \frac{1}{2}Q = Q$, he does not agree that $\frac{1}{2}Q + \frac{1}{2}Q = W$, even if W = Q. With children at Stages I and II we can admittedly explain matters by saying that, having tried to copy the volume of the whole cork to obtain a lump of clay of equal weight, they repeat the same procedure with the half or the quarter cork, because they have meanwhile forgotten that they were forced to reduce the volume of the initial ball. However, this explanation, which we had to reject even in the case of subjects at lower stages, is obviously inapplicable to subjects at Stage IIB who predict correctly that a clay ball weighing as much as the cork must be smaller than the cork. Yet despite their grasp of the relative densities of the two wholes, they never go on to divide their original clay balls into two or four when trying to copy the weight of the half or quarter corks, but behave as if the relationship between the parts were unpredictable.

Thus Kes expected that the ball corresponding in weight to the whole cork would also correspond to the half, and after having weighed it, merely removed a tenth of its substance. Ban even went so far as to produce the same (much too large) ball of clay for the half cork as he had originally offered for the whole cork; he then corrected himself but nevertheless ended up with a ball of clay larger than the one that would have corresponded in weight to the whole cork! Ale behaved even more absurdly when, in order to reproduce the weight of the half cork, he added more clay to the ball he had just balanced on the scales against the whole cork! Por stated explicitly that the two halves of the cork weigh more than the undivided cork 'because there are more pieces now'. As for the older children, e.g. Tal and Poli who, when asked to copy the weight of the half cork, immediately removed part of their first ball, they, too, did not cut it in half but only cut off a small fraction. The strange compositions produced by Kes, Tal and Poli deserve special mention. Kes, having removed a tenth of his original ball to copy the weight of the half cork, put all of it back when asked to restore the weight of the whole cork, which might have suggested that he had at least some inkling of the correct composition; however, immediately afterwards, having divided the ball into seven-eighths and one-eighth, and then into two-eighths and one-eighth, he ignored the remaining five-eighths when asked to reconstruct the weight of the whole. Tal, for his part, made the weight of the two half corks one-and-a-half times that of the whole; and Poli, having produced two balls to copy the weight of the two half corks, refused to admit that the two combined weighed as much as the undivided cork. There could be no better way of demonstrating the complete lack of quantitative composition than the reactions of these children.

How can we explain this failure? Once again it is due to logical as well as to physical factors (*cf.* §1). The reader will recall that children at Stage II fully grasp the logical basis of the quantification of matter as such. In particular, when they are handed the appropriate material, they manage, at about the age of seven years, to fuse the multiplication of relations and that of classes into a single operational whole, and hence arrive at the idea of numerical correspondence.[1] Now it is precisely the same construction which allows the child to quantify matter and that is why, as we saw in Chapter 1, the conservation of substance is the direct result of the same logico-arithmetical mechanisms as are at work at the present Stage II. However, at this level of development, the child cannot yet extend his elementary arguments from matter to weight. Thus if Q is the quantity of clay whose weight is equal to that of the whole cork (W); if Q' is the quantity of matter in a cork of weight W; and W' the weight of the clay ball of quantity Q, then the child will readily grant that $\frac{1}{2}Q + \frac{1}{2}Q = Q$, or that $4(\frac{1}{4}Q') = Q'$, etc. He will even deduce spontaneously, i.e. before any demonstration, that if $W = W'$, then $Q < Q'$ because, at equal volumes, the weight of the cork is smaller than that of the clay. From $Q < Q'$ he has no difficulty in proceeding to $\frac{1}{2}Q < \frac{1}{2}Q'$, or to $\frac{1}{4}Q' > \frac{1}{4}Q$. In other words, he realizes that Q and Q' are constant, no matter what the arrangement of their parts. By contrast, the expressions $\frac{1}{2}W + \frac{1}{2}W = W$ or $4(\frac{1}{4}W') = W'$ remain quite meaningless to him. In fact, the only way in which he can determine the weight of a given object is to balance it against another object chosen as a unit and subject to the rules of logical and numerical composition. But if the weight cannot be dissociated from the substance, the composition of such units of weight is contingent upon their being in one–one correspondence with the units of matter or the 'quantity of substance' of the object serving as the standard. Now, it is precisely this which children at Stage II fail to grasp: if W' corresponds to Q, and W to Q', they will grant that $\frac{1}{2}Q$ corresponds to $\frac{1}{2}Q'$ but not that $\frac{1}{2}Q$ corresponds to $\frac{1}{2}W$ or to $\frac{1}{2}W'$. We may therefore go so far as to say that the notion of $\frac{1}{2}W$ or W' has no meaning for the child since, when he divides Q or Q' into two parts of equal weight, he does not grasp that the sum of these two parts is equal to the weight of the initial whole. Thus Por contended that the weight of the two half corks was greater than that of the whole cork, i.e. that $W[\frac{1}{2}Q' + \frac{1}{2}Q'] > W[Q']$ because 'there are more pieces now'. Clearly, children at this stage still base all their weight determinations on purely subjective impressions. Now while this type of determination admittedly helps them to

---

[1] J. Piaget and A. Szeminska, *The Child's Conception of Number*, Routledge & Kegan Paul, 1952.

establish equivalences and differences, it in no way helps them to fit the differences into an adequate system of units. Moreover, when it comes to the construction of such a system with the help of objective equivalences they are at a complete loss.

This brings us back to our problem: how can we explain this lack of composition; or rather why does the measurement of weight not lead to a rejection of phenomenalistic egocentrism in favour of objective grouping, when this is precisely what happens with the quantity of matter? Is it for physical, logical or for physical-*cum*-logical reasons, and in the last case what is the relationship between the two? We are familiar with the physical reasons the child employs: the weight of an object is an active and substantial force depending on its structure and dimensions; weight is subject to losses during deformations and divisions, and hence unsuited to compositions of any kind. That is why Poli, having divided his ball into two unequal parts said: 'One of the two is smaller ... so it won't be as heavy (as the initial whole).' The child's logical arguments seem more mysterious: even if he goes by purely subjective impressions, we might still expect him, realizing as he does the weight of Q corresponds to that of Q', to conclude that all he has to do in order to obtain the weight of $\frac{1}{2}Q'$ is to divide Q into 2 ($\frac{1}{2}Q$). Why then does he fail to do so, and above all, even if there are physical grounds for uncertainty, why does he never even think of this solution on purely formal grounds? The question is the more perplexing in that, as we shall now see, children at Stage III discover the correct solution by a very simple method. Here, then, are some of their responses, beginning with an intermediate case whose answers do much to reveal the mechanism on which the new discovery is based.

CHIO (8; 8) is shown a cork and asked to make a lump of clay of the same weight. '*But the clay is heavier!*—So what must you do?—*Make it smaller; if I made it the same size it would weigh more.*' He makes a ball of the correct weight. 'And what about this one (the half cork)?' He removes some clay but does not weigh the remainder. 'Does the bit you have taken off weigh the same as the other half of the cork?' He weighs it in his hand. '*Yes.*—And would the two pieces together weigh as much as the whole cork?—*Not altogether, the whole cork is a little bigger.*—And what about this one (the quarter cork)?' He removes a piece from one of the reduced lumps. 'And what about the other quarter?' He removes a piece from the second of the reduced lumps.

'And what must you do to get the weight of this one (the whole cork) back?' He combines all the bits of clay with which he has been working into a ball. 'Why do you take all that?—*To see if it's the same.*—And for that one (the half cork)?' He divides his ball. 'And for the other half?—*The rest* (=the other half of the ball) *goes with this* (the other half of the cork).—And for this (the quarter cork)?' He divides half a ball into two.

OER (9; 6) makes his ball of clay *'smaller* (than the cork) *because the clay is heavier.*—And for this one (half cork)?—*I'll just halve it* (his ball). *Because this clay* (of the whole ball) *is the same as the whole cork, half the clay must be equal to half the cork.*—And what about this one (the quarter)?—*That's easy. I've only got to cut the half in two.*—And for this one (the whole cork)?—*I've only got to put the two parts* ($\frac{1}{4}+\frac{1}{4}$) *together, and then this and the other half will make the two halves, that's the whole thing all over again.'*

EST (10; 2). 'How are you going to make a ball that weighs as much as the cork?—*Smaller, because the clay is heavier.'* He does so. 'And for this one (half cork)?' He cuts his ball in two.'And is the rest equal to the other half of the cork?—*Of course it is. I've cut it in two.*—And for that one (quarter cork)?—*I'll simply cut it in two again.*—And for this one (the whole cork)?—*I just put them all together.'*

These responses seem so natural that we are surprised they did not occur at an earlier stage. Nothing could be simpler than their underlying mechanism, yet this very simplicity is based on the reversibility of thought which still eludes younger children when dealing with weights. Chio's intermediate response proved most illuminating in this respect. He began like a typical Stage II subject, by co-ordinating the weight and the volume of the ball with the weight of the whole cork, but was quite incapable of extending his co-ordination to the half or quarter corks. However, as soon as he was asked to reconstruct the whole ball he discovered that all he had to do was to combine the parts he had been working with, and this operational reversibility brought it home to him that the weight of the parts equals the weight of the whole, with the result that he could immediately proceed to the division of the whole into halves and quarters. Oer and Est, for their part, brought reversibility to bear on their constructions straightaway, as part of the very principle of composition they employed.

Must we therefore conclude that reversibility merely affects the formal mechanism of thought? Or does it result from the very elaboration of the concept of weight constituting its content?

§3. Weight and length

To answer these questions, we designed a series of special tests. Since the relationship between weight and spherical volume may have been too complex to allow of immediate quantification, we decided to present our subjects with bars of constant thickness and width but of varying lengths in order to determine whether or not they arrive at the correct proportions at an earlier stage of development than they do with the clay and cork.

We then put two questions to them. The first bore on the inverse relationship between the weight and the length: 'Will an iron bar be as long as a lead bar of equal weight?' This question was merely intended to establish whether or not the results are comparable to those obtained during the original tests. The second question bore on the quantification of weight as such: the child was asked to choose a rectangular cardboard sheet whose weight is equal to (1) a sheet of light metal of the same width and thickness; (2) half that sheet; and (3) a quarter that sheet.

Now question 1 elicited the same responses as have been described in §1: at Stage I the lengths of the sheets were equated (type 1) or the weight was said to vary as the length (type 2); at Stage II came the discovery of the inverse relationship.

Here are some examples of type 1 (Stage I):

NER (7; 0). 'Which is heavier, lead or iron?—*The lead.*—Well, look at these two bars (of which only the ends are shown). Let's say they are made of lead and iron. If they are as heavy as each other will they be the same length?—*Yes.*—What is this one made of?—*Lead.*—And that one?—*Iron.*—Will they be the same weight?—*Yes.*—And their length?—*Both the same.*—Why?—*Because they have been made to be as heavy as each other.*'

HUM (7; 0). '*The two will be the same length.*—Why?—*Because they are as heavy as each other.*'

And here two examples of type 2:

LAN (6; 0). 'Is lead heavier or lighter than iron?—*Heavier.*—Well, here is a bar of lead and a bar of iron. They are as heavy as each other. Will they be the same length?—*No. One will be longer than the other.*—Which one?—*The lead, because it's heavier.*'

LUT (7; 6). '*The lead must be longer, because lead is heavier.*—But are these two bars as heavy as each other?—*Yes.*—So which one is longer?—*The lead.*'

Now for some Stage II responses:

HER (7; 6). '*The iron bar must be longer.*—Why?—*Because iron is lighter.*—So?—*It must be longer if it is to be as heavy as the lead.*'

SCHMO (8; 9). '*The iron is longer.*—Why?—*Because iron is not as heavy as lead.*—So?—*So there must be more of it for it to be the same weight as the lead.*'

We see that these three types of reaction correspond precisely to those set out in §1, and that they can be explained in the same way.

Now for the problem of the metal sheets and the cardboard. A series of cardboard sheets of varying lengths but all of the same width and thickness are set out on the table and the child is asked to feel their weight. Once he has done so, he is handed a metal sheet of the same width and thickness and is asked to feel its weight. Next he is asked to pick out the cardboard of the same weight. Once he has chosen the right one (and after having checked the result on

175

the balance), he is asked to pick out cardboards weighing as much as a metal plate half and quarter the length of the first. Interestingly enough, the responses are precisely the same as those obtained in §2. Here, first of all, some typical Stage I responses (lengths proportional to the weights):

UDE (5; 7) selects a piece of cardboard the same length as the metal sheet, but finds that it weighs less on the balance: *'I should have taken the other one* (points to the biggest sheet, which happened to lie next to the one he has chosen).—Why?—*Because that one is smaller and this one is bigger.*—Very well.' He weighs the cardboard against the metal and discovers that they are of the same weight. 'Why do these go together?—*They don't.*—But they're both the same on the balance.—*They don't match up.*—Why not?—*The cardboard is bigger.'*

BOR (5; 3) similarly chooses a piece of cardboard the same size as the metal sheet and, after correcting his mistake, goes on to match the quarter metal sheet with a piece of cardboard the same length. When his attention is drawn to this fact, he picks up a longer strip of cardboard (first type of reaction), and then a smaller one (second type), but finishes up again with one of the same length.

During an intermediate sub-stage (Stage IIA), the child, just as he did with the cork and the clay, starts to dissociate the weight from the length, only to revert to non-differentiation once again.

HAL (6; 4). *'We need a cardboard that's not the same size because the iron is heavier.*—Well, then, choose one.' He touches several in turn, then finally selects one of the same length. 'Do you think this one will be the right weight? —*Yes, because they are the same size.'* He weighs them. *'Oh, no. The iron is a little heavier.'* He selects a larger cardboard.

He is next shown another sheet, and is given to understand by superposition (but not by explicit remarks) that it is half the size of the original one. 'Find a cardboard that weighs exactly the same.' He chooses one of the same length. 'Is that right?' He uses the balance and finds that he has been wrong. 'Try another one.' He hesitates for a long time, and again picks up one of the same length. 'Why did you do that?—*Because it's as big, so it must be as heavy.'* The correct relationship is shown to him. 'Now choose an iron sheet to go with this cardboard.' Correct. 'And for this one (the quarter)?' He picks one of the same length. *'These two are as heavy as each other.*—Why?—*Because they are as long as each other and also as wide.'*

This response highlights the difficulties children at this intermediate stage still experience, notwithstanding their initial attempt to dissociate the weight from the apparent quantity of matter. Subjects at Stage IIB, by contrast, proceed directly to the inversion of the relationship between the weight and the length when asked to match the weight of the whole metal sheet but, like the subjects described in §2—and this parallelism is highly significant—they have systematic difficulties in composing the same relationship when it comes to the half and the quarter plates.

ROS (7; 0) says that *'the cardboard is lighter'*, and immediately picks up a longer sheet to match the weight of the metal. 'And for this one (the half plate)?' Chooses a piece of cardboard a little smaller than the first. 'Why?— *Because it's a little bigger than that one* (half the metal plate).—And for this one ($\frac{1}{4}$)?' Same reaction. 'Couldn't we just take one half as long (for the half plate)?—*No, the cardboard must be a little bigger than that one* (half the metal plate).'

ZEM (8; 0) chooses a longer strip of cardboard *'because the iron is heavier* —And for this one ($\frac{1}{2}$)?—*It's been cut in two.'* But he nevertheless chooses the wrong strip. The same reaction with the quarter plate.

Here, finally, a Stage III response (correct composition of the half and quarter).

BALI (9; 11) chooses a longer cardboard strip *'because the iron is heavier.* —And for this one ($\frac{1}{2}$)?—*We must ... oh, it's a half, so we must take half the cardboard.*—And for that one ($\frac{1}{4}$)?—*That's easy, it must be half of the half, and this one ($\frac{1}{4}$ metal sheet) is half of the iron. So they'll match up.'*

We see that the correlation of weight with length runs a parallel course to the correlation of weight with volume. It follows that the difficulty in composing the weights of half and quarter the original volume or length is a systematic one, linked to the problem of quantification. We can therefore pose the problem in general terms, and, incidentally, prepare the way for Part IV.

§4. *Conclusion*

Throughout the preceding chapters we have seen that the problems posed by conservation, atomism and density are all problems in operational composition and quantification. The quantities being conserved are invariants of groupings or groups resulting from two types of reversible construction: (1) that of the whole and its parts (division), thanks to which the child discovers that the sum of the parts is equal to the whole and that it remains constant no matter how the latter is divided; and (2) that of the relations or displacements which show the child that changes in the shape of the whole are offset by compensatory changes. Now these two types of composition not only explain the genesis of the principles of conservation of matter, weight and volume at equal concentrations of matter, they are also at work in all atomistic schemata, which merely serve to take them down to the corpuscular scale. Moreover, they enable our subjects to explain expansions and contractions of bodies and also differences in density, and this thanks to that special form of displacement we have called compression and decompression. Without appeals to increases or losses in substance or weight the child thus comes to grasp the permanence of the quantity of matter and even of the corpuscular volume, and can distinguish them from the apparent quantity and volume, attributing variations in the latter to centrifugal or centripetal displacements of the elementary parts.

But what is the precise nature of these compositions; under what conditions do they first appear and how can we explain the time lag between their application to matter, weight and volume? Our attempts to retrace the child's laborious efforts to cope with density have led us straight to the problems discussed in this chapter. The great importance of the reactions we have been analysing is that they reveal a remarkable parallelism between the child's logico-arithmetical and physical arguments. We discovered — and this is an implicit corroboration of our central idea — that the problems with which we are concerned are, in fact, problems of logic as well as of physical representation. The errors made during Stages I and II reflect the child's inability to fit the parts into the whole or to invert relations, e.g. between weight and volume. What is involved here, therefore, is the structure of the logic of classes and relations or that of the arithmetical operations resulting from their fusion.

Our main problem, therefore, is to clarify the relationship between these logico-arithmetical factors and the physical concepts, or more precisely to examine the mechanism responsible for the quantification of weight and physical volume (the quantification of matter *qua* simple substance is of a piece with the genesis of quantity and number which we have described elsewhere). It is to this problem that we propose to devote the last part of this book (Chapters 10–12).

From the psychological point of view, the problem is posed as follows: when a child makes a mistake he may do so because, though he can reason correctly, he has failed to grasp the data presented to him for lack of the necessary physical concepts, or else because, though he grasps the data, he has not learned to argue formally; or for both reasons at once. In other words, the difficulties of our younger subjects may reflect a lack of thought content, a lack of formal procedures, or both, i.e. a failure to grasp the concepts of weight, physical volume and density; or a failure to fit the parts into the whole with the help of numerical composition; or both. This is an essential problem in general psychology, for its solution alone can show whether the form of thought can be dissociated from its content, and hence whether the reversibility of logico-arithmetical operations derives from those of physical operations or *vice versa*, or whether they are inseparably bound up with each other. Now from all we have gathered so far, it would seem that the form and content of thought are the less inseparable the more elementary the stage of development. The reason why the child cannot divide a weight into parts whose combination will restore the whole or quantify them into units, halves and quarters, or even invert certain relations, is not that he is incapable of performing these formal operations—he does so in other spheres with greater or lesser ease—but that these structures depend partly on the content he is asked to structure. Conversely, the content itself changes during the structuring process, which cannot therefore be derived directly from the content. There is thus a relative lack of initial differentiation between the matter to be assimilated and the schemata by which it becomes assimilated and which remain informal while they are still undifferentiated. In other words, the matter to be assimilated (the substance, weight, volume, etc.) remains irreversible while the assimilation remains imperfect, but becomes reversible upon assimilation. The result is a dissociation of a form consisting of reversible mechanisms and a content to which the form accommodates itself more or less perfectly, depending on the field of application.

From the logical point of view the problem is posed as follows: we have been able to distinguish between logico-arithmetical operations consisting of groupings of classes and relations or groups of numbers and physical operations in time and space which transform (1) class relations into divisions, (2) asymmetrical relations into displacements, and (3) numbers into measurements by which (1) and (2) can be quantified. But what is the precise link between these two types of operation? Now that we have examined the main physical operations bearing on substances, weights and volumes, we can answer this question quite easily: to that end, we need merely subject these three concepts to such logico-arithmetical operations as seriation, composition by equivalences and numerical construction. We accordingly ask the child to seriate weights and to deploy such arguments as $(A > B) + (B > C) = (A > C)$ where $>$ means heavier than (Chapter 10), or to take note of the equivalences and then deduce that $(A = B) + (B = C) = (A = C)$ or $[(A = B) + (B = C) + (C = D)] = [(A + B) = (C + D)]$ etc., both in respect of weights (Chapter 11) and also of volumes (Chapter 12).[1] If he produces the correct responses and if the logical form is independent of the physical context, it follows that such purely formal operations with the weight and volume must appear at the same time as those

[1] For the quantity of matter, see J. Piaget and A. Szeminska, *op. cit.*, Chapter X.

bearing on the quantity in general (substance), i.e. in advance of all physical operations. If, on the contrary, the content constitutes the true principle of development, the order of appearance of the physical operations will be determined exclusively by experience. If finally, as we believe, form and content are inseparable, the order of appearance of the logico-arithmetical operations will be linked to that of the physical operations, each giving rise simultaneously to a new type of logic and a new type of physics, and this according to the law of succession we have established. Moreover, if the logical and physical operations, though remaining fused, become increasingly differentiated, we shall end up with a broader logic and hence with an accelerated succession of stages on the purely formal plane, while the generalization of physical operations will remain bound up with the possibility of experimental verification. The analysis of this process, in Part IV, is certain to cast fresh light on the relationship between mental operations and experience.

PART IV

Formal Compositions

The Composition of Asymmetrical Relations and Differences in Weight

Our rational representations of the physical world involve at least three types of formal composition. We can, first of all, compose the differences between the given objects, i.e. the qualitative inequalities. This is the particular role of the logic of asymmetrical relations, which involves fitting the differences into series, or co-ordinating (multiplying) two or more seriations. We can, second, compose the qualitative equivalences; this is the role of the logic of classes (addition and multiplication of sets of equivalent terms) when no more than one propositional variable is involved, and of the logic of the symmetrical relations (equalities) when two variables are involved (these two types of logic have quasi-identical grouping structures). We can, finally, compose the equivalences and differences simultaneously, in which case we disregard the qualities, and treat individual objects as being both equal and distinct, which is tantamount to turning them into units: numbers thus result from an operational fusion of classes with asymmetrical relations.

In the final section of this book, devoted to the logic and arithmetic of weight and physical volume, we shall look, first of all, at the logic of asymmetrical relations as applied to weight differences with the help of a very simple method, namely the seriation of objects whose weights are not proportional to their volumes, so that they cannot be deduced from the latter but have to be determined by direct inspection. Moreover, lest the results of one test exert a systematic influence on the next, we conduct our several tests in random order.

Problem I

The child is shown three pebbles whose weight cannot be guessed from the volume. He is provided with a balance, and may also weigh the pebbles in his hand if he so wishes, but he is told that he is about to play a game and that one of the rules is that he must never touch more than two pebbles at a time. He is then handed two empty boxes of the same weight in which he can weigh the pebbles two at a time, and asked to put the heaviest pebble on one side, the lightest on the other, and the third in the middle. The importance of this problem resides not so much in the results it produces as in the light it throws on the serial operations employed. To that end, we make it a point to vary the tests. Thus, if a subject has established the relations $A > B$ and $A > C$, there is a 50–50 chance that he will arrange the three pebbles in the correct order ($A > B > C$) rather than in the order $A > C > B$. However, it goes without saying that in that case the conclusion does not flow from the premises. Similarly, the child may produce $A > B > C$ by pure chance, and when handed three new objects, he may simply establish $B < A$ and $C < A$, to conclude that $A > C > B$. It is therefore important to continue the tests until it becomes clear what (correct or false) system he has adopted, and how much he has left to chance. For that purpose we also ask him to solve Problems IIa and IIb.

Problem II

As a rider to Problem I, the child is invited to play two games with modelling clay. In Problem IIa he is handed three balls made of the same substance (red modelling clay), whose size varies inversely as their weight: the smallest contains some lead shot, the second-smallest a pebble, and the largest nothing but clay. The child is told that the three balls do not have the weight they seem to have, and that he must therefore weigh them two at a time before putting them in the correct order.

With Problem IIb he is again handed three balls of different sizes, but this time the heaviest is that of medium size, and the lightest the biggest. The instructions are the same.

Problem III

The child is asked to arrange four to six pebbles whose difference in weight he cannot establish by mere inspection (IIIa), or three pebbles of identical volume but of different weight (IIIb). He is again told to serialise them by weighing two of them at a time.

Problem IV

Once the first three problems have been solved, we go on to compare the results to those of a general problem in seriation. The child is shown ten clay balls of the same volume but of different weights, and is asked to arrange them in order of increasing weight, without any special rules. This means that he no longer has to weigh them two at a time.

Problem V

The child is handed three matchboxes (same brand and no other distinguishing marks), and is told that they are of unequal weight, the heaviest being full of sand, the middle one full of matches and the lightest empty. He is asked to weigh them in his hand. The boxes are then shuffled around, and put back on the table to form a triangle, and the child is asked to answer the following questions by pointing to the right boxes, without opening or touching any of them: Problem Va: This box (A) is heavier than that one (B), and that one (B) is heavier than this one (C). Which is the heaviest of the three boxes? And which is the lightest? Problem Vb: This box (A) is heavier than that one (B), and this one (C) is lighter than that one (B). Which is the heaviest of the three? And which is the lightest? Problem Vc: This box (B) is lighter than that one (A), and heavier than this one (C). Which is the heaviest and which is the lightest of the three?

The results of these tests are highly instructive. At Stage I, the child cannot solve Problems I and II because he weighs only two of the three objects, often one at a time, i.e. without any correlation. He fails *a fortiori* with Problems III (four stones) and V (verbal questions), nor can he solve Problem IV (simple seriation). At Stage II, as well, he fails to solve Problems I and II, but this time because he establishes his relations by co-ordinating isolated pairs, e.g. $A > B$ and $A > C$. Similarly, with Problem III, he simply establishes that $A > B$ and $C > D$, and fails to appreciate that this tells him nothing about the relationship between A, B and C, D. When it comes to the simple seriations (Problem IV), he proceeds empirically, starting with pairs or triplets, but fails to co-ordinate his successive constructions. He also fails to provide the correct answers to Problem V. At Stage IIIB, finally, he constructs the correct series $A > B > C$ by co-ordinating all the relations (Problems I and II), and also produces the correct seriation (Problem IV). However, at the beginning of this stage (Sub-stage IIIA) he still fails to construct the logical system needed to solve Problem III, and though he successfully composes the simple relations Va and Vb, he fails to co-ordinate the inverse relations ($B < A$ and $B > C$) (Problem Vc). In general, Sub-stage IIIA thus marks the discovery of operational seriation, which is completed at Sub-stage IIIB.

§1. *Stage I: Lack of composition*

When four- to five-year-olds are asked to seriate three pebbles by weight two at a time (Problem I), they often fail to solve the problem, for the simple reason that they see no point in weighing all three pebbles. Instead they content themselves with weighing just two, and deciding whether they are heavy or light in themselves. Here are a few examples:

BUR (5; 8) places stone A on one of the balance pans and stone C on the other. He discovers that C is lighter and puts it down to his right. To the left of C he puts down A as the heavier of the two and then adds B to the front of his series without having weighed it, as if it were self-evident that it must be the heaviest. 'But have you weighed this one (B)? – *It's heavier.* – Why? – *Because that one* (C) *is light, and this one* (A) *is a little heavier.'*

SIN (5; 10) weighs A and B on the balance but does not really grasp the function of this instrument. He is asked to feel the weight of the pebbles and puts down $A \leftarrow B$[1]. Then he puts down C to the right: 'Have you weighed it? – *No.* – Where should it go?' He picks it up again, feels it in his hand and says: *'It's light.'* Whence $A \leftarrow B \leftarrow C$, but by pure chance. After another test (IV) he is handed back the three pebbles. He weighs A and C in his hand and puts down $A \leftarrow C$, then puts down B in front of A without weighing it: 'Why did you put it there? – *This one* (B) *is the heavy one.'* He is then allowed to touch all three pebbles, and seriates them correctly, but fails with more than three or four pebbles (Problem IV).

Next he is presented with Problem IIa: seriation of three balls of clay whose weight is in inverse ratio to their volume. He weighs A and B but nevertheless considers B heavier because it is 'bigger'. Next, he weighs C in his hand and places it between the other two, whence $B \leftarrow C \leftarrow A$. 'How do you know that (C) is heavier than (A)? – *Because it is heavy.'*

BED (5; 11) weighs B against C and puts down $B \leftarrow C$. Next he puts down A to the right of C, whence $B \leftarrow C \leftarrow A$. 'Why? – *Because it is light.'*

Problem IIb: He weighs and puts down $A \leftarrow C$ and continues his series with B. 'Why have you put (B) over there? – *That's where it goes.'*

We see that at this stage co-ordination is impeded by misjudgment of the relations. The child is quite capable – and this from a very tender age – of comparing two weights and hence of establishing the correct perceptive relation $U \rightarrow W$ or $V \leftarrow W$. Apart from the illusions engendered by Problem II – as we know the weight and the apparent quantity of matter are not dissociated until Stage II – he therefore has no difficulties in making perceptive comparisons of the weights of two elements. Indeed such comparisons are inherent in the perception of weight as such and we know, from Köhler's work on the colour perception of chicks, that all perception – however elementary – is based on relations and not on

---

[1] $A \leftarrow B$ means A is heavier than B, and is equivalent to $B \rightarrow A$.

absolute qualities. In other words the perception of qualities and of the relations between them is a primitive faculty, the relations providing the child with a rough idea of quantity (in contrast to intensive quantification based on seriation and extensive quantification based on the numerical unit), while quality still remains in the crude state. However, these perceptive relations, precisely because they are not yet quantifiable (extensively or intensively) or even seriable, are not yet logical relations: the asymmetrical relation only appears with the appropriate grouping, i.e. with the seriation or co-ordination of the relations themselves.

Thus, even while they declare that A is heavier than B, children at this stage still fail to grasp the relativity of weight and express the perceptive relation between two objects by calling one 'heavy' and the other one 'light'. Thus, Sin placed B in front of A because he thought it 'a heavy one'; and even when these children use such comparisons as 'heavier', 'lighter', etc., they are still thinking in terms of perceptive rather than operational relations. This is borne out by their treatment of the third pebble: the reason why they do not weigh it is that they are quite incapable of co-ordinating two separate relations, in other words of constructing a series of three terms. What they are trying to establish instead is a set of binary relations, i.e. something halfway between classification and seriation: they seek out the 'heavy ones' and the 'light ones', and when they put the third stone in front or at the end of the other two, they do so simply to place it on the heavy or on the light side. Thus, when Sin nevertheless placed C between B and A, he immediately added that C was heavy. Since he obviously felt no need to weigh the third pebble against either of the other two, he could not possibly have been trying to construct a series, but was simply trying to establish whether that pebble was heavy or light. This view is corroborated by the peculiar weighing method employed by children at this stage: if they do not immediately put the third pebble down, they will often weigh it carefully, but by itself; which proves conclusively that they are simply trying to qualify it as light or heavy. Others, again, weigh all three stones separately, which compels them to compare the first two. Moreover, even when they weigh the pebbles in their hands, they do not aim at correlation, but simply try to separate the 'heavy' from the 'light' (Sin).

It is therefore only natural that when these children are presented with Problem III (four stones), they should fare no better.

SIN (5; 10) picks up A and B and puts A to his right because 'it is the heaviest', and B to the far left of A. He then picks up C, weighs it and inserts it between the other two, but nearer to B. Finally he weighs D and places it in the empty space between A and C, whence the purely fortuitous series $B \leftarrow C \leftarrow D \leftarrow A$.

NAR (5; 11) weighs A and C and puts down C←A. Next he weighs D in his hand and places it in front of C; then he weighs B and places it after A, whence D←C←A←B.

Problem V (boxes) elicits similar responses:

COL (5; 10). 'You see these boxes, etc. This one (A) is heavier than that one (B), and that one (B) is heavier than this one (C). Would you repeat what I have just said? – *This one is heavier than that one, and that one is heavier than this one* (i.e. A←B←C). – Now put the heaviest of the three over here, the lightest there, and the third one in the middle.' He puts down A←C←B. 'Are you sure? – *No, this one* (B←C←A) *is better*.'

These reactions are just what we might have expected from the solutions of Problems I–III. Again, when it comes to simple seriation (Problem IV) the reactions are comparable to those associated with the construction of lengths and quantities of matter: our subjects fail to produce the correct series of ten balls and merely construct a kind of global arrangement in which most of the heavy balls are on one side, and the light ones on the other. However, this type of construction persists much longer in the case of weights than it does with simple quantities (lengths or breadths).

ALB (5; 11). 'You see these balls here (in disorder). You might think they all weigh the same, but they are all different. Could you put the heaviest of all over here (to the right), and the one that's just a little lighter next to it, then the one that is a little lighter still, etc. and the lightest of all over there (to the left).' Alb picks up ball 6 which is right in front of him, weighs it by itself and puts it down; next he picks up ball 8 and puts it down to the right of ball 6; followed by ball 3, which he says is 'light' and which he puts down to the far left of ball 6; ball 9, which he says is 'heavy' and which he puts to the left of 6; ball 7, which he puts down to the left of 9; ball 5, which he puts down to the left of 7; ball 4, which he calls 'light' and puts down to the left of 3. Next he picks up ball 10 and exclaims 'Gosh, it's heavy', and puts it down between 6 and 8. Finally he weighs the two remaining balls in his hands, says that 'they're both of them light', and puts them right in front of the rest, whence the series: 2←1←4←3←5←7←9←6←10←8.

The original instructions are repeated: 'Make sure you've done it right, the heaviest of all must be over here, the next heaviest by its side, etc.' Alb touches the balls one by one but does nothing to change his arrangement. The demonstrator then places ball 10 in front of ball 1, but to no effect.

Alb's responses are so characteristic that we need quote no others.[1] They show that when the series is long enough and the child has complete freedom of manipulation, he will succeed neither in ordering the terms nor in adding the relations. Again, when there are only three or four pebbles he may admittedly prove capable of

[1] For the case of lengths and breadths, see J. Piaget and A. Szeminska, *The Child's Conception of Number*, Routledge & Kegan Paul, 1952, Chapter V.

seriation, but only if he is not given the restrictive instruction to handle no more than two objects at a time.

Thus Sin produced the correct series of three pebbles without, but not with, the restriction. This type of response is the result of perceptive comparisons, not of deductive reasoning. The child either picks up two pebbles in one hand and the third pebble in the other, or else he weighs all three in one hand in quick succession and recalls their weights, the near-simultaneous memory impressions left by a small enough number of objects being almost the same as their simultaneous perception. If we took a film of these weighings (of three or four pebbles with complete freedom of manipulation) and if we then played the film back in slow motion, we should undoubtedly see that the child performs the very operations in practice that are needed for the solution of Problems I, II, and III, and which he fails to obtain by reflection when he is instructed not to touch more than two objects at a time. However, the main difference between the direct and unconscious pre-operations our slow-motion film would reveal and the indirect operations the child must perform to solve Problems I–III is that in the former, the immediate tactilo-motor memory stands in for reason, while in the latter the child must co-ordinate the relations he has perceived by a series of judgments, and hence rely no longer on his memory but on a system of reference points, e.g. one in which the heaviest object is placed to the right and the lightest to the left in such a way that he can always tell in what order the next comparison must be made. If we fail to distinguish pseudo-seriations based on the unconscious memory or on direct perception from true operational seriations we might easily conclude that Köhler's chicks were able to solve the very problem that baffled our children when, after choosing the darker of two colours $A \rightarrow B$, they went on to peck at an even darker colour, C. Now it is obvious that these poor beasts do not construct the series $A \rightarrow B \rightarrow C$ but that they rely on their perceptive memory, once they have been trained to ignore the brighter of two colours. In brief, real seriation does not begin until intellectual co-ordination is brought to bear on the successive relations.

Now, when our subjects are asked to seriate ten elements rather than three or four, with complete liberty of manipulation, the problem is complicated once again, and becomes as hard to solve as the seriation of three elements taken two at a time. This is because it is no longer possible to touch all the elements simultaneously or to weigh them in the hand in quick succession and then to record the perceptive relations in the memory. To seriate ten elements, the child must be able to establish a law of succession and grasp that each term must be both lighter than all the preceding and heavier than all the successive terms. This construction therefore demands the correlation of $(A \leftarrow B)$ with $(B \leftarrow C)$, and this is why Problem IV proves as hard to solve at Stage I as do Problems I and II. Needless to say the detailed solutions are not completely synchronous; there are some children who are better at seriating ten elements, because of special experiences, than they are at co-ordinating two isolated relations, and *vice versa*. But, by and large, these two types of behaviour may be said to be homologous.

We have dwelt at such length on the difference between real and pseudo-seriations because this distinction is crucial to the comparison of all three methods of quantification considered in this book (substance, weight and volume). It is obvious that, if extensive quantification does indeed result from the operational combination of seriations and equivalences, as we have supposed it to do, then the child will find it much easier to quantify directly visible qualities than any others: the seriation (and the determination of equivalences) of direct visual data can range over a very much larger field of simultaneous perception. This is why the quantification, the seriation and the equalization of the apparent quantities of matter are constructed first: these quantities are open to direct visual inspection, while the construction of the conservation of weight and physical volume is bound up with either a non-visual perceptive field or else with data that, though supplied by the eye, have nothing direct about them and call for a complex intellectual elaboration. Hence the importance of a detailed analysis of seriation and of the other quantifying operations to anyone anxious to explain the time lag between the constructions of the conservation of substance, weight and physical volume. We shall return to this problem at the end of this chapter, but we have thought it appropriate to mention it at the start of our study of seriation, if only to drive home the fact that it is because of these psychological difficulties that the same logical groupings are elaborated at distinct intellectual levels, depending on the perceptive or experimental contents to which they are applied.

§2. *Stage II: empirical seriation by unco-ordinated pairs*

Stage II brings a striking advance: the child now weighs each of the elements against another, having realised that the weighing of an isolated element tells him nothing. However, this advance does not automatically lead to the solution of Problems I–III because, in addition to weighing the elements two at a time, the resulting pairs must be co-ordinated if they are to constitute a series. Thus, while the relations $A \leftarrow B$ and $B \leftarrow C$ can be added together, no clear conclusions can be drawn from $A \leftarrow B$ and $A \leftarrow C$, and nothing at all can be deduced from $A \leftarrow B$ and $C \leftarrow D$ unless C is compared with B. Seriation therefore demands that the correlated pairs interlock, and moreover, in a fixed order. In the event, however, we see that children at Stage II try to solve Problems I and II by no other weighings than $(A \leftarrow B)$ and $(A \leftarrow C)$, and Problem III by constructing juxtaposed couples in isolation.

Here, first of all, are responses to Problems I and II:

PER (6; 10) begins, like children at Stage I, by weighing A against C, and then puts B in the third place without having compared it to either. But a week later he weighs A against B, then A against C and constructs $A \leftarrow C \leftarrow B$. 'How do you know that this one (B) is lighter? – *I saw it from that one* (A).'

Problem IIa. He establishes $A \leftarrow B$ then $C \leftarrow A$ and puts B after C, whence $A \leftarrow B \leftarrow C$: 'How do you know that (C) is the lightest? – *Oh, it should be like this* $(A \leftarrow C \leftarrow B)$.'

Problem IIb. He establishes $B \rightarrow A$, $C \rightarrow A$, and constructs $A \leftarrow B \leftarrow C$. 'Is this one (B) heavier than that one (C)? – *Yes.* – Why? – *Because it comes first.*' In other words he is mistaking his arbitrary arrangement for proof.

REY (7; 8). Problem I. He establishes $B \rightarrow A$, $C \rightarrow A$, and places C in the centre, whence $B \rightarrow C \rightarrow A$. 'Which is the lighter, this one (C) or that one (B)? – ... – Why did you put this one (C) in the middle? – *Because it's lighter than that one* (A). – And this one (B)? – *The same.* – Well, look, here are three other pebbles. Try to do better'. He establishes $A \leftarrow C$, $B \leftarrow C$, and puts down $B \leftarrow A \leftarrow C$. 'Which is the heaviest? – *That one* (B). – Why? – *It is heavier than this one* (C). – And that one (A)? – ...'

Problem IIa. He establishes $A \leftarrow B$, $A \leftarrow C$, and puts down $A \leftarrow C \leftarrow B$. 'Which is the lightest? – *This one* (B). – Why?' He weighs B against C and produces the correct series. But with Problem IIb, he reverts to his original mistake.

MOR (7; 10). Problem I. He establishes $B \leftarrow C$, $B \leftarrow A$, and correctly puts down $A \leftarrow B \leftarrow C$, but purely by chance as his subsequent reactions show.

Problem IIa. He establishes $A \leftarrow B$ and $A \leftarrow C$ and puts down $A \leftarrow C \leftarrow B$. 'Which is the lightest? – *That one* (B). – Why? – *Because I saw it weighed less than this one* (A). – And what about that one (B)? – *It's a little heavier.*'

The importance of these responses only becomes clear when we compare them to the equivalent reactions to the four-pebble test (Problem III).

MOR (7; 10) establishes $C \leftarrow D$, and puts down $D \rightarrow C$. Next he weighs A against B and puts down $A \leftarrow B$, but to the right of the first pair, thus ending up with $D \leftarrow C \leftarrow A \leftarrow B$. 'Which is the lightest of the four? – *That one* (B). – How do you know? – *It's lighter than this one* (A). – And which of these two (A and C) is the heavier? – *That one* (C). – How do you know? – *I've weighed it.* – Which is the heaviest of all four? – *That one* (B). – Are you sure?' He picks up C and D and corrects his error, then checks A against B. '*Here you are* $(C \leftarrow D \leftarrow A \leftarrow B)$.'

ORA (7; 11) establishes $A \leftarrow C$, $B \leftarrow D$, and puts down $A \leftarrow C \leftarrow B \leftarrow D$. 'Which is heavier, this one (C) or that one (B)? – *That one* (C). – Why? – *Because* (= it's a fact).'

BER (7; 11) establishes $C \leftarrow D$, $A \leftarrow B$, and puts down $C \leftarrow A \leftarrow D \leftarrow B$. 'Which is the heaviest of the four? – *That one* (C). – How do you know that it's heavier than this one (A)? – *Because it comes in front* (!)'

We thus have a beginning of correlation: the child is no longer content to divide the perceptive elements into heavy and light ones, thus treating relations as absolute qualities. In particular, he no longer tries to establish the weight of an isolated element, and from the fact that A is heavier than C, he no longer concludes that B is bound to be light. However, his nascent relativity stops short because he juxtaposes rather than co-ordinates the relations he has established. There is, therefore, no proper system of relationships, but only a system of pre-relative qualities.

This pre-relativity is a constant feature of all these reactions to Problems I and II. Thus, if chance leads him to $A \leftarrow B$ and $A \leftarrow C$ rather than to $A \leftarrow B$ and $B \leftarrow C$, he is perfectly content and puts down ACB or ABC indifferently, i.e. he simply juxtaposes the two relations $A \leftarrow B$ and $A \leftarrow C$. More significant still is his justification of his arbitrary seriations. He may, first of all, revert to qualification; thus, when Per argued that B is lighter than C because 'I saw it from that one (A)', he was simply saying that since B proved to be lighter than A it must perforce be lighter than C. In other words B is no longer 'lighter' in itself, as it was at Stage I, but thanks to the kind of generalization characteristic of what we have called pre-relations. It was in this sense, too, that Mor thought B lighter as such, simply because he had found it lighter than A. Alternatively the child at this stage treats the series he himself has constructed as objectively true. Thus Ber claimed that B was heavier than C 'because it comes in front'.

The reactions to Problem III bear out this interpretation to the full. Either the four elements are simply juxtaposed – $(C \leftarrow D) \leftarrow (A \leftarrow B)$ (Mor), or $(A \leftarrow C) \leftarrow (B \leftarrow D)$ (Ora) – or else there is an attempt at synthesis, as with Ber, and we obtain the most typical of all pre-relations: because C is 'heavier' than D, and A is 'heavier' than B, C and A are placed at the head of the series as being 'heavier' in themselves, while D and B come at the end as being 'lighter' in themselves, whence the series $C \leftarrow A \leftarrow D \leftarrow B$. The reason why C is heavier than A, according to Ber and Per, was simply that 'it comes in front'.

It goes without saying that, under these conditions, no semi-formal reasoning is possible, whence the failure to solve Problem V:

CUE (7; 2). 'Look at these three boxes (A ← B and B ← C). Which is the heaviest of the three? – *That one* (A). – And the lightest of them? – *That one* (B). – And if I take these two (B and C), which would you say is lighter? – *This one* (C). – And if I were to take all three which is the lightest? – ...'

In respect of the general seriation (Problem IV), we see that as soon as the child ceases to offer a global solution (Stage I) and begins to evaluate the terms, he also tends to proceed by pairs or small global series of three or four terms. At the early phases of this stage he leaves it at that, but at later phases he goes on to adjust successive elements empirically until, finally, he succeeds in regularizing the series by simple trial and error.

MIC (6; 0) arranges his balls as follows: 1 ← 3; 2 ← 4; 5 ← 6; 7 ← 8; 10 ← 9: except for the last relation, the series consists of unco-ordinated pairs of a heavy element followed by a light one. When asked to check his construction, Mic picks up two elements at a time, and produces the correct series.

BAR (7; 0) constructs the pairs: 1 ← 2; 4 ← 3; 5 ← 7; 6 ← 9; 8 ← 10; and then corrects the order of succession of the terms 9 and 8.

DAL (7; 5) begins with 1 ← 3; 2 ← 4; then puts down 6 ← 7 ← 8 and 9 ← 10. Next he interposes the term 5 between 4 and 6, checks the whole series, and corrects the beginning.

By and large, therefore, these children proceed from the heavier to the lighter balls, but empirically, i.e. without a rigorous system in which every ball is both lighter than the preceding and heavier than all the remaining ones. At Stage II, moreover, they find simple seriation somewhat easier than solving Problems I–III. This suggests that the free manipulation of the material which, as we saw at the end of §1, lends itself directly to the practical seriation of small sets of three or four terms, can at a given moment turn into a more comprehensive method, while the reflective analysis of the relations begins with the method of pairs. The seriations characteristic of Stage II are therefore a mixture of these two methods in varied proportions.

§3. *Stage III: operational seriation*

At Stage III, the seriation of weight is completed, though in two phases. At Sub-stage IIIA serial additions are still confined to two one-way relations (Problems I, II, Va and Vb), and there is also a grasp of free seriation (Problem IV), but the more complex relations (Problems III and Vc) are not constructed until Sub-stage IIIB.

Let us look first at some Stage III responses, beginning with two transitional cases:

DUT (9; 0). Problem I: he establishes A ← B and A ← C, and puts down

$A \leftarrow C \leftarrow B$. 'How do you know that C is heavier than B? – *Because I've weighed it* $(A \leftarrow B)$. – What did you find? – *Oh, I've forgotten to weigh the rest.*' He weighs B against C and puts down the correct series $A \leftarrow B \leftarrow C$.

Problem III: he establishes $A \leftarrow B$, puts down A to his left and B to his extreme right, then establishes $C \leftarrow D$ and puts them in the centre, thus obtaining $A \leftarrow C \leftarrow D \leftarrow B$.

SPA (10; 0). Problem I: he establishes $A \leftarrow B$, $A \leftarrow C$, reflects, establishes $B \leftarrow C$, and puts down $A \leftarrow B \leftarrow C$.

Problem IIa: he weighs $A \leftarrow C$, $A \leftarrow B$, $B \leftarrow C$, then puts down $A \leftarrow C \leftarrow B$. 'Are you sure?' He checks $A \leftarrow B$, $B \leftarrow C$, and puts down $A \leftarrow B \leftarrow C$.

Problem III: he weighs $C \leftarrow D$, $A \leftarrow B$ and puts down $C \leftarrow A \leftarrow D \leftarrow B$. 'Is this one (D) heavier than that one (B)?' He checks. *'No.'* He puts down $C \leftarrow A \leftarrow B \leftarrow D$. 'And the rest?' He weighs them two at a time and finishes up with $A \leftarrow C \leftarrow B \leftarrow D$.

Problem IIIb (four balls): he weighs $A \leftarrow B$, $B \leftarrow C$, $C \leftarrow D$, and ends up with $A \leftarrow B \leftarrow D \leftarrow C$.

POR (6; 8, advanced). Problem I: he weighs $A \leftarrow B$ and $B \leftarrow C$, whence $A \leftarrow B \leftarrow C$.

Problem IIa: he weighs $A \leftarrow B$ and $A \leftarrow C$, puts down $A \leftarrow C \leftarrow B$, but spontaneously checks $B \leftarrow C$, whence $A \leftarrow B \leftarrow C$.

Problem III: he weighs $A \leftarrow B$, $B \leftarrow C$, $C \leftarrow D$, but puts down $A \leftarrow B \leftarrow D \leftarrow C$.

Problem IIIb: he weighs $A \leftarrow B$, $D \leftarrow B$, $D \leftarrow C$, and again puts down $A \leftarrow B \leftarrow D \leftarrow C$.

JER (8; 9). Problem I: he weighs $A \leftarrow B$ and $A \leftarrow C$, but, just before constructing the final series, checks $B \leftarrow C$ and puts down $A \leftarrow B \leftarrow C$.

Problem IIa: he weighs $A \leftarrow B$ and $B \leftarrow C$, whence $A \leftarrow B \leftarrow C$.

Problem III: he weighs $A \leftarrow B$, $C \leftarrow D$, $A \leftarrow C$, and finally $B \leftarrow D$ and puts down $A \leftarrow B \leftarrow C \leftarrow D$. But this conclusion, though correct, is not justified, since $B \leftarrow C$ is missing.

Problem IIIb: he weighs $A \leftarrow B$, $A \leftarrow C$, $C \leftarrow D$, and $B \leftarrow D$, and again puts down $A \leftarrow B \leftarrow C \leftarrow D$ (same lack of $B \leftarrow C$).

NEM (9; 10). Problem I: he weighs $A \leftarrow C$ and $A \leftarrow B$, checks $B \leftarrow C$, and goes on to construct the correct series.

Problem II: he weighs $B \leftarrow C$ and $A \leftarrow B$, whence $A \leftarrow B \leftarrow C$.

Problem III: he weighs $B \leftarrow C$, $C \leftarrow D$, $A \leftarrow D$, $B \leftarrow D$, $A \leftarrow B$, and puts down $A \leftarrow B \leftarrow D \leftarrow C$, thus forgetting $C \leftarrow D$.

We see how much progress has been made since the last stage in respect of Problems I and II. Apart from a few residual reactions (e.g. Dut's) all these subjects have come to realize that, in addition to establishing $A \leftarrow B$ and $A \leftarrow C$, they must also establish $B \leftarrow C$, i.e. co-ordinate all the terms and not simply juxtapose successive relations. There is also marked progress with Problem III. Except for the two transitional cases, these subjects no longer content themselves with two comparative weighings, but make at least three. In other words they have realized that the seriation of four elements calls for more than the construction of two isolated pairs.

They nevertheless still make mistakes, though often in minor respects only. Thus Por, who had established all the relevant data (A ← B, B ← C and C ← D), failed twice to reach the correct solution because of what we might call an error in calculation. Ger arrived at the solution by chance, but could not deduce it as a necessity because, though he established two relations that were not needed, he forgot one that was. Nem, who had constructed all the relations plus two redundant ones, forgot the second one and thus also finished up with a calculating error. Now, no matter whether the child constructs a relation and forgets it, or whether he fails to construct it in the first instance, he commits a methodical error: instead of proceeding in an orderly fashion and consistently choosing the heaviest (or the lightest) element of every pair, he leaves the choice to chance, and hence fails to correlate the relations he has established.

By contrast, subjects at this stage are successful with Problem IV (free manipulation of the material) though the age at which they produce a correct solution may vary by several months, depending on the nature of the elements presented to them. Thus Binet and Simon[1] have shown that when children are asked to seriate five boxes weighing 3, 6, 9, 12 and 15 grammes respectively, and are given three minutes to do so, they do not achieve a 75 per cent success rate before the age of ten years. With the ten balls of 100–250 grammes in geometrical progression (Weber's law) which we have used, without a time limit, the operational seriation appears at the average age of nine years, i.e. at Sub-stage IIIA. For example:

PAT (8; 11) chooses the heaviest of the ten balls, then, holding it in his hand, seeks out the heaviest of the rest, puts down the first and repeats the same procedure with the second, etc.

Now it is this very procedure which subjects at Sub-stage IIIB employ to solve Problem III: what we have here is an interesting age shift to which we shall be returning.

Now, for some reactions to the semi-verbal questions (Problem V):

POR (6; 8). Problem Va: 'This box (A) is heavier than that one (B), and this one (C) is lighter than (B). Which is the heaviest of the three? – *That one* (C). – And the lightest? – *This one* (A)' (Correct).
    Problem Vb: 'This box (A), etc. . . . Which is the lightest of the three? – *That one* (C). – And the heaviest? – *This one* (A).'
    Problem Vc: 'This box (B) is heavier than that one (C), and at the same time it's lighter than this one (A). Which is the heaviest of the three? – *That*

[1] A. Binet and T. Simon, *La Mesure du développement de l'intelligence chez les jeunes enfants*, Paris, 1929, pp. 20–2.

one (B). – Why? – *Because it's heavier than that one* (C).'

JAC (8; 10). Problem Vb: 'This box, etc. (i.e. A ← B and C ← B). Which is the heaviest? – *That one* (A). – And the lightest? – *That one* (C).'

Problem Vc: 'This box, etc. (i.e. B ← A and B ← C). Which is the heaviest? – (B). – Why? – ... – And which is lighter, A or C? – (A).'

In brief, when it comes to two one-way relations such as A ← B and B ← C (Problem IVa), or to a single inversion of the type A ← C and C → B (Problem Vb), the child at this stage quickly arrives at the correct solution, which calls for no greater effort than the seriation of the three pebbles. But when he has to cope with two inverse relations, such as B → A and B ← C, the difficulty is much greater and the child reverts to prerelations.[1]

Let us finally look at Sub-stage IIIB responses, i.e. at the correct solutions of all the problems:

JUN (7; 3). Problem I: he weighs A ← B and B ← C, and puts down A ← B ← C but checks A ← C: 'Why are you weighing these? – *To find the heaviest.*'

Problem IIa: he weighs A ← B and B ← C, and puts down A ← B ← C straightaway.

Problem III: he weighs A ← B, B ← C and C ← D, and puts down D to the left. He continues with B ← C and A ← C and puts C to the right of D, then proceeds with A ← B and puts down B followed by A, whence the series A ← B ← C ← D by successive eliminations of the lighter element.

Problem IIIb: he weighs B ← C and keeps B, then B ← D and A ← B, and puts down A. He continues with B ← D and B ← C and puts down B, then weighs C ← D, whence A ← B ← C ← D, by successive eliminations of the heavier element.

Problem IVc: immediate seriation.

Problem Vb: A ← B and C ← B; points out the lightest and the heaviest correctly.

Problem Vc: correct responses.

MEY (9; 5). Problems I and II: correct answer without hesitation.

Problem III: weighs A ← B, A ← C and A ← D, and puts A to the right; continues with C → B and B → D, and puts B to the left of A; follows with C ← D, whence D → C → B → A.

Problem IIIb: 'Try to do it with the least possible number of weighings. – *Yes, like this* (A ← B, B ← C, B ← D).' He puts down A ← B ← D ← C. 'Are you sure?' Correction.

Problem IV: immediate seriation.

Problems Va, Vb, and Vc: correct responses.

MONT (10; 3). Problem I: weighs C → A and puts C to the left while holding A in one hand and comparing it with B, whence B → A; puts A to the right and continues with C, whence B ← C; and concludes with A ← B ← C.

Problem II: same method.

Problem III: he weighs B ← C, B → D and A ← B; continues with A ← C,

[1] See our article 'Une Forme verbale de la comparaison chez l'enfant', *Archives de Psychologie*, 1921, 20.

A ← D, and puts A down on the right. He picks up B, checks B ← C and B ← D, and puts B to the left of A. He weighs C ← D and puts down D → C → B → A.

Problem IIIb: 'Very good, but now try to do it with the least number of weighings.' He weighs A ← B and C ← D. 'These here (A and C) are the heaviest and these (B and D) are the lightest. − Very well.' He weighs A ← C and is about to put down A ← C ← B ← D, but decides to check C against B, whence A ← B ← C ← D.

Problems IV and V: correct.

Since Problems I and II are solved by children at Sub-stage IIIA, we need not return to them, except to note Jun's reaction; he started out with an inductive approach but eventually resorted to pure deduction.

As for Problem III, there is a remarkable parallelism between its solution at Sub-stage IIIB and that of Problem IV (free seriation of the ten elements) at Sub-stage IIIA. In both cases the subject tries systematically to discover the heaviest (or the lightest) of all the elements, next the heaviest (or lightest) of the remainder, etc. There is, in fact, nothing astonishing about this parallelism because seriation is merely the generalization of the co-ordinative operations (addition of asymmetrical relations) involved in Problems I–III. Nevertheless, free seriations prove slightly in advance of shorter seriations with the instruction to weigh two objects at a time. Thus, at Stage I the former is still as hard as the solution of Problems I and II; at Stage II, it marks a slight advance, and at Stage III it is a whole sub-stage ahead of the solution of Problem III. Now this apparently complex development is quite simple to explain. Because at Stage I the seriation of the ten elements exceeds the limits of perception and of the immediate memory, and also because the child still lacks an inductive and operational method, he succeeds no better with Problem IV than he does with Problems I, II, and III. At Stage II, though he has not yet reached the operational level, he has learned to proceed by intuitive induction, and can therefore seriate the ten elements by trial and error and empirical co-ordination, but still fails with Problems I–III because here the co-ordination of pairs calls for a deductive and operational method. At Sub-stage IIIA he becomes able to co-ordinate the relations operationally. He therefore finds it as easy to seriate ten elements by successively selecting the heaviest (or lightest) as to co-ordinate the (two) relations involved in Problems I and II. By contrast, when it comes to the addition of three relations (Problem III) it does not occur to him to apply the method he has used for the free seriation of the ten terms, because the latter presupposes a prior order of succession, whereas in the former he is instructed to weigh the elements two at a time and believes that he can link what relations he has established in any way he pleases. Now, while he often succeeds with two relations (or three terms), he is generally at a loss with three relations (or four terms); as witness the reactions of Por, Jer, and Nem who forgot what they had, and had not yet, done. At Sub-stage IIIB he keeps all the relevant data in his mind and returns to the method which has stood him in such good stead at Sub-stage IIIA for the solution of Problem IV. Moreover, when he is asked to produce the correct solution with a minimum number of weighings, he often succeeds completely or nearly so (e.g. Mont).

Most of our subjects were also successful with the semi-verbal questions (Problem V), although Problem Vc still caused them some difficulty (*cf.* Jun's reaction).

§4. *Conclusions*

What strikes us first of all about these seriations is their complete convergence with the seriations of lengths, heights, widths, etc. In an earlier study of the child's conception of number (with Mlle Szeminska) we also had occasion to introduce qualitative seriations: we asked our subjects to seriate rods, clay balls, etc. Moreover, though we did not pose Problems I–III in the form presented here (which would have had no sense since all the data were open to visual inspection), we also looked at serial and ordinal correspondence, i.e. at the qualitative or numerical correlation of two similar series. Now we were able to establish precisely the same three stages of development as those we have just described: global arrangements without seriation, followed by empirical seriations, and finally by operational seriations leading to the coordination of all the relations. However, and this is the chief importance of this comparison, the development was completed at about the age of six or seven years, i.e. well ahead of the seriation of weights. In particular, the seriation of ten clay balls whose weight is proportional to the apparent volume comes earlier than that of balls of the same volume but of different weights (Problem IV).

It is therefore legitimate to assume that the seriation of quantities of matter precedes that of weights, when the latter are not proportional to the former. Moreover, when the volumes differ markedly, they, too, can be seriated with relative ease, but if they do not, and their seriation calls for indirect methods, e.g. the displacement of liquids, it goes without saying that their seriation will prove even more difficult than that of weights.

The logic of asymmetrical relations involved in seriation is, in fact, a formal mechanism which, once he has discovered it in one sphere, the child might be expected to extend quite quickly to the rest. But if that is so, why do subjects capable of seriating rods or balls of different sizes fail to apply the same method straight away to the seriation of weights? The problem is the more important as it is quite general. It concerns not only the physical operations leading to the successive construction of the three invariants (substance, weight and volume) discussed in Chapters 1–6, but also the purely logico-arithmetical operations we shall be examining in Chapters 11 and 12; where we shall see that the simplest equivalences $(A = A'; A' = A''; \text{ hence } A = A'')$, or the most elementary additive compositions are applied to the quantity of matter well before they are applied to the weight, and to the weight well before they are applied to the volume. Moreover, as we shall see in Chapter 11, the composition of weight equivalences runs parallel to the seriation of unequal weights: when the child is unable to seriate two unequal relations, he will also fail to derive one equivalence from two others, and as he reaches the operational level in one of these fields he also reaches it in the other. The problem of the time-lag between the construction of the three constants (substance, weight and volume) thus concerns all formal or logico-arithmetical operations and not merely those involved in seriation. How are we to resolve this problem?

We must begin by separating two questions that, though they eventually become one, are quite distinct at first: that of the perceptive conditions responsible for the beginnings of seriation, and that of the intellectual conditions responsible for operationality. The first can be put as follows: why is it easier to compare differences or equivalences perceived by the eye than it is to compare weights felt in the hand, and why are weights easier to compare than volumes? The second question can also be rephrased: why does the child, having discovered the grouping of the first of our three qualities, not apply that grouping immediately to the second and third, and why do we find the same differences in their logical composition as appear in their perceptive structuring?

Now, as far as the first question is concerned, it is easy to show that the apparent quantity of matter always manifests itself in the form of lengths, heights, widths, denumerable sets, etc., i.e. in the form of data that visual perception can normally combine into a single whole, and whose seriation and correlation it facilitates in any case. Whenever the weight of two or more objects corresponds to their visible dimensions, the subject might well resort to intuitive seriation, were it not that experience teaches him that the weight is not always proportional to the quantity of matter. Now, when a series of objects differs in weight, but not visibly so, as in the present test, the perceptive field constituted by the tactile determinations is much narrower than the visual field; hence the co-ordination of the perceptive relations calls for a much greater number of intellectual operations. And when it comes to volumes, it is clear that direct visual inspection will again enable the child to evaluate them when the objects are of similar shape or when they differ appreciably in their dimensions. However, when the differences or equivalences are not apparent at first sight — and this is precisely what happens during deformations of the clay balls — or, in general, whenever there are marked differences in shape, then perception is not enough, and the child is forced to resort to indirect methods of evaluation, e.g. the displacement of liquids. The very conception of volume thus involves intellectual operations and is no longer based on perceptive qualities, whence the delay in its logical treatment and, indeed, in its very discovery and differentiation from matter and weight.

However, appeal to the perceptive factors can only help us to solve the first of our two questions: it merely explains the time-lag between seriations involving matter, weight and volume respectively, but does not tell us why, once the child has constructed the operational mechanism in the first sphere (for example when he discovers that, to seriate rods, he must always select the longest of those left behind and that, in case of difficulty, he need merely pick up his last selection and compare it with the remaining rods one at a time), he may take up to two years to apply the same mechanism to the weight, though he could easily repair his lack of visual perception by means of manual weighings. And above all, once he has adopted the operational mechanism, why does he never even think of determining the objective relations by reference to the balance? The reason, as we shall see in the next chapter, is that the balance in no way helps the child to establish the inference $(A = A') + (A' = A'') = (A = A'')$, and this simply because he refuses to accept it. Must we therefore take it that the time-lag between the perceptive evaluation of weight and volume cannot be made good and hence prevails at all stages of mental development, or must we look for a deeper explanation?

Now, what is the real cause of the child's difficulty in fitting a quality into a system of asymmetrical relations or of equivalences, in seriating it, and composing it in various ways, and in quantifying it? As we have shown at length, in respect of physical operations and *a fortiori* in the sphere of formal or logical operations, the chief obstacle to grouping is egocentrism, i.e. co-ordination based exclusively on unreflective perception. The opposite of logical composition is not chaos, for no thought at all is possible without a reference system. The reason why the child fails to group one relation with all the other possible ones into a coherent whole is that he refers it to an absolute construction outside the grouping. Now, this absolute, which to the child resides quite naturally in the object, cannot be grasped except by reference to the self. The reason why a stone is not 'light' in the precise sense that it is both lighter than a well-ordered series of heavy stones and heavier than a well-ordered series of lighter stones is that it is 'light' in itself; but what is the meaning of 'light in itself' if not light with respect to the subject? This is the true significance of intellectual egocentrism and, if we employ this term in the sense we have given it, it is obvious that egocentrism is the contrary of grouping. Now, it is this very phenomenon which we observe during the first two stages described in this chapter (with residual traces up to Sub-stage IIIA): when the child simply divides his balls into light and heavy ones, when he combines A with C as being the 'heaviest' without bothering to check if A is heavier than B or if C is heavier than D, in brief when he qualifies rather than seriates, his reference system is obviously not yet based on reversible compositions but on qualities informed by his own actions.

But, if egocentrism is thus the antithesis of grouping, are we not being tautological when we explain the existence of the first by the absence of the second? This would be so if it were not possible to determine the intensity of the factors involved. In fact, precisely because of the perceptive conditions we have just described, it is easy to show why weight remains an egocentric quality much longer than the visible quality of the quantity of matter. Vision, and particularly his visual space, confront the subject with a universe of simultaneous data to which he at first responds with a radical form of egocentrism, but which he objectifies during the first two years of his life by grouping the displacements he perceives and constructing the invariant object. Moreover, except for relations strongly dependent on the position of his own body, such as left and right or his general perspective, the child quickly learns that 'large' and 'small' and 'broad' and 'narrow', etc. do not exist as such, and therefore subjects the visual data to logico-arithmetical operations. However, because 'heavy' and 'light' are more difficult to structure perceptively there is not only a time-lag, but also a fixation of egocentric habits, which render these qualities more resistant to relativization, i.e. to equalization and measurement. As for physical volume, which is wholly relative, it goes without saying that its logical composition will come much later still.

These, then, are the two reasons which explain the time-lag between the formal composition of the quantity of matter, of weight and of volume. But there is more to come. We do not reason logically except about invariants of thought. Can these logical invariants be constructed even while the corresponding physical invariants are still lacking? And what is the precise relationship between them? It is to this problem that we shall be devoting the next chapter.

CHAPTER ELEVEN

Simple and Additive Compositions of Equivalent Weights[1]

Having examined the composition of unequal weights which falls in the province of the logic of asymmetrical relations, we must now look at simple and additive co-ordinations of equivalences which fall in the province of the logic of symmetrical relations and of classes if the equivalences are qualitative, and in the province of numbers and measurable quantities if they are iterated.

We shall thus come back to the central problem of this entire work. In Part I we saw that the child finds it very hard to grasp that the whole remains equal to the sum of its parts no matter what their arrangement. In Part II the problem of atomistic composition was shown to be beset by the same difficulties. Then in Part III we saw that compositions of weights and volumes involving the summation of the parts are anything but self-evident to the child, a point we must now examine at greater length.

When we try to elicit the purely formal structure (the logico-arithmetical operations) of reasoning by reducing the physical problems to a minimum, i.e. by reducing the parts of the whole to the smallest possible number and also by making the qualitative differences as elementary as possible, shall we find that the child concludes that A has the same weight as C once he has established that $A = B$ and $B = C$? And will he be able to arrive at $(A + B) = (C + D)$ if he himself has discovered that $A = B = C = D$?

To settle these points, we present our subjects with a series of brass bars of the same length and thickness but of different widths, which can be composed by simple juxtapositions and superpositions in the following manner: (1) four narrow bars (Ia, Ib, Ic and Id) are equivalent to a plate IV; (2) any three of these bars are equivalent to the smaller plate III, so that $III + I = IV$; (3) two smaller plates still, IIa and IIb, are of such width that $IIa + IIb = IV$, or $IIa + Ia + Ib = IV$, or $IIa + Ia = III$, each II being equivalent to two bars of the order I. Invariably, $Ia = Ib = Ic = Id$. In addition we

[1] In collaboration with Mme Ingold-Favroz-Coune.

also show the child a piece of lead, a carbon plate, a lump of dry wax, a lump of iron and a ball of clay (perhaps constructed by the child himself) all equal in weight to a brass bar of order I. The brass bars are either left plain or else they are painted white, black, blue, red and orange respectively.

The children are first asked to weigh the bars so as to make sure that $Ia = Ib$, and that one of them is equal in weight to the piece of lead, the carbon plate, etc. They are then asked to make homogeneous compositions of several brass objects or heterogeneous compositions of objects differing in density. The comr ositions of the first type will be said to be simple when they are confined to the co-ordination of equivalences,[1] e.g. $(Ia = Ib) + (Ib = Ic) = (Ia = Ic)$, or additive when they involve the (logical or numerical) summation of the parts, e.g. $(Ia + Ib = IIa)$ or $(III + Ic = IV)$, etc. In the second case, the objects are placed in the two pans of a balance, which is left on the table for the child's use. The heterogeneous compositions, too, may be simple or additive. Here is an example of what may be asked about the first type: 'If Ia has the same weight as the piece of lead (as determined on the balance), and if $Ia = Ib$ (also determined on the balance), will Ib have the same weight as the piece of lead?' Next, for a question relating to the second type: the demonstrator places $(Ia + Ib)$ in one pan and $(Ic + \text{lead})$ in the other, and asks the child if the weights in the two pans will be equivalent or not.

As a result of such questions we were able to establish (1) that at Stage I, children are quite incapable of producing simple or additive compositions even with homogeneous objects. The only apparent exception was the equivalence $Ia = Ib = Ic$, etc. when the brass bars are left plain; but this is obviously not a logical composition, because as soon as we paint the bars we find that the child fails to deduce $Ia = Ic$ from $Ia = Ib$ and $Ib = Ic$; (2) that at Stage II the child produces simple and additive compositions of homogeneous objects but goes wrong (or needs many trials and errors) with simple compositions of several homogeneous objects and one heterogeneous one; and fails completely in the case of additive composition of heterogeneous objects; (3) that, at Stage III finally, he succeeds with all the heterogeneous compositions, both simple and additive. However, we must once again distinguish an intermediate Sub-stage IIIA, during which the child begins to make correct deductions with simple heterogeneous compositions and also succeeds by trial and error with the additive composition of heterogeneous objects; at Sub-stage IIIB he effects all possible compositions by means of a fully deductive method. Quite obviously, the extensive or metric quantification of weight, too, does not

[1] By addition of the relations themselves, but not of their terms.

appear until Stage III, when it can be effected thanks to the fusion of the composition of equivalences with seriations; during Stage II, by contrast, only the quantity of matter can be subjected to numerical or metric composition, and at Stage I there is no (extensive) quantitative composition of any kind.

§1. *Stage I : no composition*

At this, the lowest stage, which continues until the age of six or seven years, the child proves incapable of effecting compositions even in respect of homogeneous objects. He may, nevertheless, produce the correct answer, for instance when he identifies the unpainted bars Ia, Ib and Ic arranged side by side, or when he identifies three bars with plate III on which they are resting without overlap; but as soon as the bars are separated from each other or from plate III, he will deny their equivalence. In other words he still relies exclusively on perceptive or sensori-motor assimilation. Here are some examples:

GAI (5; 4) fails to compose the weights even by the simple addition of halves. Thus, when shown two identical corks, one of which is cut in two, he says: '*The whole cork will be heavier* (than the two halves). — Why? — *Because it's thicker.*'

He is next shown the brass plate III on one pan and the bars IIa and Ia on the other: 'Will they be as heavy as each other? — *No, this lot* (IIa + Ia) *will be heavier.* — Why? — *Because they're in two parts.* — And if we put these (IIa + Ia) on top of that one (III) will they be as big?' He does so. '*Yes.* — And on the balance? — *That one* (III) *will be heavier.* — Why? — *Because it's bigger.* — But how did we get these two (IIa and Ia)? — *We took a piece like that one* (III) *and cut it.* — So, are they as heavy? — *No, this one* (III) *will be heavier because it's bigger.*'

He is next shown two bars, Ia and Ib, which he predicts will be of the same weight: 'And on the balance? — *They'll weigh the same.* — And these? (Ia and Pb[1])? — *They probably won't be the same weight.* — Try it.' He uses the balance. '*Oh yes, they are.* — And what about these (Ib and Pb)? — *This one* (Pb) *is heavier because it's bigger.*' He weighs it in the hand. 'And on the balance? — *It's heavier as well.*' He weighs it. '*Oh no, it's the same weight.*'

Once convinced by a series of weighings that Pb = Ia, Ib or Ic, Gai is shown Ia + Ib in one pan and Ic + Pb in the other. '*That lot* (Ia + Ib) *will be heavier.* — But what would happen if we weighed this one (Ia) against that one (Pb) on the balance? — *They'd weigh the same.* — And what about these (Ia + Ib), and those (Ic + Pb)? — *The second lot will be heavier because the lead is thicker.*'

He is next shown plate III on one pan and IIa + Pb on the other. 'Will they be as heavy as each other? — *No, this one* (III) *will be heavier because it's*

[1] Abbreviations used: Pb = the piece of lead; C = the carbon plate; CL = the ball of clay; all of the same weight as I.

bigger. *Oh no, this lot will be heavier because it's got two different ones* (IIa + Pb).'

COL (5; 10) is shown III on one pan and Ia + Ib + Ic on the other: 'Are they the same weight? — *No, this one* (III) *is heavier because it's bigger.* — Let's see.' The three Is are placed on top of III, and Col runs his fingers over them to make sure they fit precisely, then they are returned to the pan. 'Well? — *The bigger one* (III) *will be heavier.*'

He is next shown Ia + Pb: 'Are they the same weight? — *No, this one* (Pb) *is heavier because it's bigger.* — See for yourself.' He weighs them. *'They're the same.* — Why? — *Because the lead isn't so big after all.* — And these two (Ia = Ib)? — *Oh, they're like that* (stretching gesture). — And these (Ia and Pb)? — *We saw they're the same.* — And these (Ib and Pb)? — *This one* (Ib) *is heavier and the square one* (Pb) *is lighter.*' Ia and Ib are put on one side of the pan, and Ic + Pb on the other: 'Will they balance? — *No.* — Why not? — *This side* (Ia + Ib) *is heavier because there are two of the same, and those* (Ic + Pb) *aren't.* — Do you remember these two (Pb and Ia)? — *Yes, they're the same.* — So what about these (Ia + Ib and Ic + Pb)? — *The first two are heavier because they're long ones.*' Pb is weighed against Ia, Ib and Ic in turn: 'Are they all the same (two at a time)? — *Yes, they are.* — So what about these (Ia + Ib) and (Ic + Pb)? — *The second lot are lighter.* — How can you tell? — *Because I counted them, over here there is only one* (Ic) *and over there* (Ia + Ib), *there are two of the same.* — And what about this one (Pb)? — *It's not the same.*'

VIR (6; 1). Unpainted bars: 'Does this one (Ia) weigh as much as that one (Ib)? — *Yes.* — And what about these (Ib and Ic)? — *The same.*' He presses them against each other. 'What are you doing? — *I'm looking to see if they're the same.* — And these (Ia and Ic)? — *The same.*' We start all over again: 'What about these (Ia and Ib)? — *The same.* — And these (Ib and Ic)? — *The same.* — And these (Ia and Ic placed at right angles to each other)? — *No.* — Why not? — *This one* (Ia) *is heavier.*'

With III and (Ia + Ib + Ic) Vir picks up the three bars and places them on top of III: '*If we take them three at a time they weigh the same.* — And like this (separated)? — *That one* (III) *is heavier than these* (Ia + Ib + Ic). — Why? — *Because it's bigger.*'

Tests with the lead produce the same type of response. For (Ia + Ib) and (Ic + Pb) Vir says: *'These two* (Ia + Ib) *are heavier, there are two of them and over here* (Ic + Pb) *we've got that lump.*'

PAS (6; 3). Painted bars: 'Does the red one (Ia) weigh the same as the blue one (Ib)? — *I'll have to feel them in my hands.*' He does so. *'Yes.* — What about the orange (Ic) and the blue (Ib)?' He weighs them in his hand *'The same.* — And the orange (Ic) and the red (Ia)?' He tries to weigh them. 'No, guess; what do you think? — ...'

Same result with Pb: he establishes that Pb = Ia, and recalls that Ia = Ib. 'And what about the lead and the blue (Ib)? — *The lead is heavier.*'

BOUG (6; 10) weighs the red against the orange and says: *'They balance.* — And if we change the red for the blue?' He does so. *'It's the same.* — Well, what if we weigh the red against the blue? — *The blue will be heavier.*'

Having established the equality of the three bars, he is asked: 'What about the lead and the blue? — *The lead'll be heavier, because it's bigger.*' He weighs them. *'No, they're the same.* — What about the red? — *The lead will be heavier.* — See for yourself.' He weighs them. *'No, they're the same.* — And what about the orange? — *The lead will be heavier.* — Make sure!' He weighs them. *'Oh no, they're the same as well.'*

(Ia + Pb) and (Ib + Ic): *'The red and the lead are heavier.* — Why? — *Because lead is heavy.'*

These typical Stage I reactions reflect a complete failure to compose weights even by simple equivalence or addition of homogeneous parts.

In fact, even those subjects who have established the identity of the dimensions by the superposition of homogeneous objects, e.g. Ia + Ib + Ic = III, or III = IIa + Ia (and we begin with these subjects because they bear out our predictions in Chapters 2 and 9), do not infer the equivalence of their weights. Thus Gai, having said that IIa + Ia are as big as III, went on to declare that, when they are no longer superposed, III becomes heavier 'because it's bigger'. Vir, who spontaneously placed Ia + Ib + Ic on top of III and said 'If we take them three at a time they weigh the same', also went on to say that, as soon as they are separated, III becomes heavier because it is 'bigger'.

Moreover, these subjects cannot even apply the most elementary logical composition to the case of the bars I, i.e. they do not conclude from A = B and B = C that A = C. Admittedly, when they are shown the two unpainted and parallel bars Ia and Ib they can identify them by direct visual inspection, and also the bars Ib and Ic, and if they are then asked if Ia = Ic, they will give the correct answer, but in that case, their three judgments Ia = Ib, Ib = Ic and Ia = Ic are purely perceptive evaluations. We have only to substitute two painted bars and to arrange them slantwise, for these subjects to deny that Ia = Ic! Thus, Vir, immediately after declaring that Ia = Ib = Ic when the bars were parallel, also granted that Ia = Ib and Ib = Ic, but denied the equality when Ic was placed at right angles to Ia. As for Pas and Boug, as soon as the bars were painted three different colours they, too, refused to conclude from Ia = Ib and Ib = Ic that Ia = Ic.

Needless to say, these children are even less capable of deducing A = C from A = B and B = C when confronted with the lead. Thus Boug kept saying that the lead would be heavier than any of the bars he was shown, and this after the balance had belied his prediction time and again. Moreover, when Col and Gai discovered that Ia = Ib, which they had predicted, and next that Ia = Pb, which they had not, and were then asked if the lead would be as heavy as Ib they forgot everything that had gone before, and failed to draw the right conclusions.

Finally, even when the lead has been found to weigh as much as every one of the bars, and when the child no longer has any doubts in this respect, it suffices to introduce the more complex system (Ia + Ib) and (Ic + Pb) for him to deny what he has only just granted. Thus Col declared that (Ia + Ib) would be heavier 'because they are two things the same', as if their qualitative resemblance turned them into two units, unlike the lead and the other bar which were 'not the same'. Vir also said that (Ia + Ib) were heavier because 'there are two of them', and that (Ic + Pb) were lighter because they included 'that lump'. Gai, for his part, thought that (IIa + Pb) were heavier than III 'because it's got two different ones', which agreed with his initial view that the two halves of a cork weigh more than the whole, but also shows that he did not treat 'one' and 'two' as real numbers but as intuitive units not subject to composition.

Now, all these reactions raise the problem of the relationship between metric quantification, or numerical composition, and the composition of simple or additive logical equivalences. A number is, in fact, part of a system of equivalent but distinct units, such that, when they are separated in any order, the resulting series will all be similar (vicarious order). Now a continuous quantity (such as weight) constitutes an analogous system of units, but applied to variations of a specific quality; or, which comes back to the same thing, it is a system of asymmetrical relations such that $a + a' = b$, $b + b' = c$, $c + c' = d$, etc. (where a, b, c, d . . . represent the increasing differences between the first and successive items, and a', b', c' . . . the differences between the second and third; the third and the fourth terms, etc., but in such manner that a' is equal to a; b' is equal to a'; c' is equal to b', etc.).

Number or measurable quantities thus involve (1) the grouping of equivalences, (2) the seriation of the terms or relations, and (3) the operational combination of these two groupings into a single one (numerical or metric composition). Are these three conditions satisfied at the stage under review?

We just saw that condition (1), i.e. the grouping of the equivalences $(1 = B, B = C, hence A = C)$, is definitely not. Now, these Stage I constructions of equivalences correspond to Stage I constructions of seriations, and we saw in the last chapter that the composition of asymmetrical relations eludes children at this stage no less than does the composition of equivalences. In fact, in the same way that the subjects whose responses we have just mentioned cannot add and co-ordinate the equivalences $(A = B)$ and $(B = C)$ into a single relation $(A = C)$, and this because they consider every relation in perceptive isolation rather than logically, so they also cannot seriate A, B, C, because having discovered that A ← B, they see no need for correlating C with either term, and simply endow it with a non-relative perceptive quality. In both cases, the relations they establish are therefore intransitive and not open to composition. It goes without saying that, in these circumstances, there can be no operational combination of equivalences and asymmetrical relations, and this is precisely why children at Stage I cannot effect numerical compositions or extensive quantifications of weights. Weight is not yet a quantum, but a simple subjective quality whose true measure is still the subjective impression it produces on the hand.

Let us note, moreover, that this situation is not peculiar to the case of weight, and that, at Stage I, substance, too, cannot be quantified by means of extensive units. This is what we were able to establish in an earlier work, in which we dealt with the decantation of liquids, and in which we saw that if the contents of a vessel A are poured into a vessel B and from B into C the child fails to deduce $B + C = 2A$ from $A = B$ and $B = C$.[1]

In any case, it is clear that the homogeneous compositions of the brass bars and plates concern the quality of matter no less than the weight, because the density of these objects is the same, so that their weight is proportional to their substance.

In other words, these elementary reactions are of a kind with the Stage I reactions we have discussed in previous chapters, all of which revealed a failure to quantify weight and volume, or even to grasp the very simple idea of the conservation of substance. Now, there is nothing astounding about this similarity when we remember that children at this stage are quite incapable of grasping, on the purely logical level, that the whole is equal to the sum of its parts, that two quantities equal to a third are equal to each other, or that three non-equivalent terms can be seriated if the differences between the first and the second, and between the second and the third, are known.

§2. Stage II : correct homogeneous compositions but failure with simple or additive heterogeneous compositions

Children at Stage II are capable of constructing series of simple equivalences $(A = B) + (B = C) = (A = C)$, or of additive equivalences $(Ia + Ib + Ic = III$, etc.) when the objects presented to them are of the same material (bars or plates of brass), but they fail

[1] J. Piaget and A. Szeminska, *The Child's Conception of Number*, Routledge & Kegan Paul, 1952, Chapter X. It should be noted that the present Stage I corresponds to Stages I and II in the construction of number.

to effect the same compositions when an object of different density is introduced (heterogeneous composition). In other words, they are able to quantify matter (and consequently the weight if it is proportional to the apparent quantity of matter), but they cannot yet produce logical compositions or quantifications of weights not proportional to the apparent quantity of matter. Here are some examples, starting with an intermediate case:

MAN (5; 4) predicts spontaneously that Ia = Ib = Ic = Id, but when Ia and Ib are placed parallel to each other on one pan and Ic and Id are placed at right angles on the other, he hesitates: *'These two* (Ia + Ib) *are heavier, or perhaps they weigh the same.'* Later, after tests we shall describe below, he is asked: 'Does this one (III) weigh as much as these (Ia + Ib + Ic)?—*They weigh the same.'*

By contrast, having first declared that Ia and the lead *'weigh the same, because the lead is smaller',* and having checked the correctness of his prediction, he nevertheless denies that Ib = Pb. 'Why not?—*Because that one* (Ib) *is made of iron, and so it's heavier.'* For (Ia + Ib) and (Ic + Pb) he says: *'The first lot will weigh more; the two long bits are heavier. The second lot is lighter because it's got one long one and one short one.'* But a moment later he claims that (Ic + Pb) are heavier *'because of that square'* (the piece of lead).

CLA (6; 6) establishes the equality of the blue and the black bars, and of the black and the red bars. 'Which ones do we still have to weigh?—*The red and the blue. They will be the same as well.—*Why?—*Because one is over here* (the blue on the left), *and the other over there* (the red on the right), *and because...'* He moves the black bar from one pan to the other, thus showing that if A = C and B = C then B must ensure the equality of A and C.

But despite his certainty, Cla fails to apply the same operational schema to the composition of the weights of heterogeneous objects: 'Is this piece of lead as heavy as the black bar?—*No, the bar is heavier, because it's longer.'* He weighs them. *'Oh, they're the same.—*And the black bar and the red?—*They're the same as well.—*And the lead and the red?—*The bar is longer so it'll be heavier.'*

Having established that Pb is equal to any one bar of order I, the demonstrator produces (Ia + Ib) and (Ic + Pb): 'Will they still weigh the same?—*No, this lot* (Ic + Pb) *is heavier: it's got one thick one and one long one.'*

GRI (6; 10). 'Does the red bar weigh as much as the blue?—*Yes.—*Why?—*They are the same size.—*And the blue and the black?—*The same.—*And the red and the black (arranged perpendicularly)?—*They're still the same weight.*

'And the lead and this red bar?—*The lead is heavier because it's a bit thicker.* —Weigh it!—*Oh no, it's the same.—*And this one (the blue bar) and the red?—*The same.—*And the lead and the blue bar?—*This one* (the bar) *is lighter.—* Why?—*Because the lead is heavier.'*

BOL (7; 0) also concedes that (Ia = Ib), (Ib = Ic) and (Ia = Ic), but fails with the lead and pebble test.

When a pebble is placed on one pan and a bar on the other, he weighs them and exclaims, *'Gosh, it's the same!—*Now, listen. I'm going to add a bar to

210

each side. Will they still balance?—*The two bars are the same, I saw that they're as big as each other, and these two are the same as well* (the pebble and the first bar) *but the pebble is thicker.*—So?—*This side* (pebble plus bar) *will be heavier because there is a bar and this heavy one* (the pebble).'

When Bol has discovered that the lead is equal in weight to each of the bars, the demonstrator places Ia in one pan and Pb in the other. 'Are they the same weight?—*Yes.*—Now watch, I am adding a bar to each side (Ib and Ic), whence (Ia + Ic) and (Ic + Pb). What about their weight?—*The one with the lead will be heavier.'*

ROD (8; 0). 'What about the red and the blue bars?' He weighs them. '*They're the same.*—And the red and the black?—*As well.*—So?—*So all three are the same.*—And the lead and the blue bar?—*The lead is heavier because it's not made of iron.*—Weigh them.—*They're the same.*—And the lead and the red?—*The lead is heavier.*—Why?—*It's not made of iron.*—And the blue and the red?—*They're the same.*—What happened when you weighed the red with the blue?—*They were the same.*—And the lead with the red?—*The lead was heavier.*—Weigh them!—*Oh, they're the same.*—And the lead with the black?—*They're the same; oh no, the lead is heavier than the iron.*—Make sure!' He weighs them. '*Oh, they're the same again.*—And what about these (Pb + red) against those (black + blue)?—*This side* (Pb + red) *will be heavier because lead is heavier than iron.'*

These Stage II reactions reflect a marked advance over the preceding ones: while subjects at Stage I fail to produce any kind of deductive composition, the present subjects immediately deduce (A = C) from (A = B) and (B = C) when dealing with homogeneous bars of different colour and arrangement, and also prove successful with such additions as Ia + Ib = IIa, etc.

Thus Cla 'and Rod realized that, because the red and the blue were equal to the black they must also be equal to each other. Similarly Rod, having seen that A = B and B = C, immediately concluded that all three were 'the same'. But significant though this advance from perceptive assimilation to deduction undoubtedly is, we must remember that the bars are identical except for their colour and position. In other words, we remain close to the sphere of perception; the composition can still be based on the shape and the dimensions abstracted from other qualities, and hence only bears on the quantity of matter and on weights proportional to it.

As for composition by addition, our subjects, e.g. Man, immediately equate III with Ia + Ib + Ic, etc., which distinguishes them from subjects at Stage I. But, here again, though there is marked progress we must not exaggerate its importance: it bears essentially on the dimensions, and hence on the quantity of matter.

But what happens when the weight is no longer proportional to the quantity of matter and when the child has to establish simple or additive equivalences of the weights as such, i.e. of the weights

211

of objects of different densities? At Stage II this type of composition still eludes him, even when he himself has established successive equivalences by consulting the balance.

In fact, as soon as we substitute the lead or pebble or any other heterogeneous object for one of the bars, these children refuse to use the same process of reasoning they have just applied to the bars. The responses of Cla and Rod are again particularly interesting, since, though both had established the operational schema $(A = B) + (B = C) = (A = C)$, they obviously failed to extend this schema to the new situation. Thus, immediately after they were shown that the lead weighs precisely as much as one of the bars, they kept insisting that it was heavier. Man and Gri reacted in the same way, but Gri also made it clear why he considered hetero-geneous compositions quite unlike homogeneous compositions: the reason why bar A = bar C, bar A = bar B, and bar B = bar C is that they are 'the same size', whereas, once the lead is introduced, the dimensions are no longer comparable: 'the lead is ... a bit thicker'. Clearly, therefore, these compositions of the weights of the bars are merely compositions of quantities of matter; only those involving heterogeneous objects are true compositions of weights. Now, it might be argued that the failure to produce the latter is due to a memory lapse: the child expects that the lead will always be heavier and hence forgets the evidence of his own eyes. However, Rod's response ('they're the same ... oh no, the lead is heavier than the iron') shows that what we have here is not forgetfulness but a conflict between logic and intuition which these subjects resolve by allowing subjectivity to take precedence over quantifying logical composition.

As for additive compositions of heterogeneous objects, it goes without saying that the child at this stage fails with them completely. Even after he has established the equivalence of the lead to every one of the bars, the lead has merely to be presented in a new system of four units, for him to assert once again that it is heavier (or lighter). Moreover, in some cases, we have simplified the test by adding one bar to each pan just after the child has seen that the two pans balance, but even then Bol, for instance, refused to admit the equivalence. Thus when the pebble (A) and the bar (B) were balanced with two bars (C and D), he stated explicitly that A = C and B = D, but claimed that $(A + B) \leftarrow (C + D)$: 'The two bars are the same and these two are the same as well ... so this side will be heavier.' Greater disdain of logico-arithmetical composition can hardly be imagined.

In short, while children at Stage II can effect compositions with homogeneous objects based on the co-ordination of simple equivalences and the addition of equivalent terms, they fail to construct the same compositions with heterogeneous objects. The reason for this difference, as we saw, is that the first are, in fact, compositions of quantities of matter, while compositions of the second type alone are true compositions of weight. These conclusions provide a striking corroboration of those we arrived at in Chapter 9: when children at Stage II try to shape a clay ball of the same weight as a large cork, they produce a smaller construction because they have learned to disassociate the weight from the quantity of matter, but when it comes to reproducing the weight of half or quarter the cork, they do not cut their balls in two or four because they have not yet learned to compose or to quantify weights.

This distinction has a fundamental bearing on the general problem of quantification: at Stage II matter has become quantifiable, while the quantification of weight has not progressed over what it was at Stage I. If, as we have assumed in our earlier study of the construction of number, metric quantification is indeed the consequence of an operational synthesis of seriations and equivalences (or of asymmetrical relations and classes), this difference is easy to explain. In respect of the quantity of matter the child at this stage has learned to seriate magnitudes, so that when we present him with a system of vessels and liquids he solves all the problems in much the same way as he solves those posed in the present test by the bars and the plates.[1] In other words, the synthesis of these ordinations and equalities leads him to number and to the measurement of quantities in general, and of the quantity of matter in particular. By contrast, just as he still fails to correlate the weights themselves (heterogeneous composition), so he also still fails to seriate unequal weights (Chapter 10), and this for the same reason: in both cases he succeeds in establishing such perceptive relations as $A = B$; $B = C$; $C = D$, etc. where $A \leftarrow B$; $A \leftarrow C$; $C \leftarrow D$, etc., but he fails to co-ordinate them operationally into $(A + B) = (C = D)$ or into $(A \leftarrow B \leftarrow C \leftarrow D)$ because he proceeds by juxtaposed or mixed pairs and does not even deduce $(A = B)$ from $(A = C)$ and $(B = C)$, or manage to seriate two relations in the order $A \leftarrow B$ and $B \leftarrow C$ rather than $A \leftarrow B$ and $A \leftarrow C$. Now, if children at Stage II can neither seriate weights nor compose them transitively they can quite obviously not quantify them metrically; they continue to treat weights as egocentric and phenomenalistic qualities. For that reason they also fail to treat individual objects as weight units in heterogeneous compositions. Thus, Cla, who realized that the bars were interchangeable units, refused to equate $(Ia + Ib)$ with $(Ic + Pb)$ because 'this lot ... has got one thick one

[1] J. Piaget and A. Szeminska, *op. cit.*, Chapter X.

and one long one', i.e. because the lead and the brass Ic cannot be treated as two comparable units.

This failure to quantify weights coupled to the successful composition of the quantity of matter is thus of a piece with Stage II responses to simple conservation (Chs 1 and 2), to corpuscular conservation (Chs 4 and 7), and to the relationship between weight and substance in objects of different densities (Chs 8 and 9).

§3. *Stage III: heterogeneous compositions and the intensive and metric quantification of weight*

At Stage III, finally, the operational schemata of simple and additive equivalence are applied to the weight as such. However, this advance proceeds by two steps, so that we must once again distinguish two sub-stages. During the first (Sub-stage IIIA), the subject arrives at the correct solution by a semi-intuitive and semi-operational method, reflected in a series of trials and errors. True, when two of the three terms in the simple composition $(A = B) + (B = C) = (A = C)$ are homogeneous, the child at this stage will produce the correct deduction straightaway, but only because he is helped by the intuitive equivalence of two of the terms. By contrast, when all three terms of the simple composition are heterogeneous, or when the additive composition involves at least one heterogeneous term, he will still proceed by trial and error. During Sub-stage IIIB, by contrast, all his compositions are deductive from the outset. Here then are some Sub-stage IIIA responses, beginning with an intermediate case:

CHA (6; 10) grasps the equality of the bars immediately after pressing two of them together: *'That makes that* (superposition). *You can see for yourself.—* And these two (Pb and Ia)?—*The lead is heavier.* (Weighing.) *Oh no, they're the same.—*And these (Ib and Pb)?—*The bar is heavier.* (Weighing.) *I don't know why, but they're the same as well.—*And these (Pb and Ic)?—*Oh, they're the same because this bar and that are the same, just look!—*And these (Pb and Id)?' He hesitates. *'They're the same, I saw they are, because this one* (Id) *is the same as these two* (Ib and Ic).—What if I put these two (Pb + Ic) in one pan and those two (Ia + Ib) in the other?—*This side* (Pb + Ic) *will be heavier, because the lead is heavier.*' He is shown Pb and the bar separately. *'Ah, they're the same.'* The four objects are put back on the scales: *'These* (the two rods) *are heavier, because here there are two and over there it's just one.'*

MUR (6; 11). '(Pb and Ia)?—*I don't know.*—Weigh them.—*They're the same.*—And this one (Pb) with that (Ib)?—*It's the same as all of them, because all the bars weigh the same.*—And (Pb + Ia) against (Ib + Ic)?—*This lot* (Pb + Ia) *is heavier because the lead is thicker.'* Eventually he grants: *'they're the same'*.

BOD (7; 0). 'Does this piece of lead weigh as much as that blue bar?—(Weighing.) *Yes.*—And the red and the black?—*The same.*—And what are we left with?—*The blue, but they're all the same. We saw that they were with the black and the red.*—What about the blue and the lead?—*The lead will be heavier, but it's almost the same as the iron.*—Weigh them.—*Oh no, they weigh exactly the same.*—And the lead and the red?—*They're the same as well.*—Why?—*Because we saw on the scales that the blue and the lead are the same, and before that we saw that the red and the blue and the black are the same.*

'What if I put the blue and the red in one pan, and the lead and the black in the other?—*The lead and the black will be heavier because the lead has more weight than the black bar. Oh no, they're the same, but the two bars will still be lighter.*—And the lead with one of these bars (the blue)?—*It's the same.*—And if I put the lead and the blue in one pan, and the black and the red in the other?—*If we turned them into balls they would be the same; the lead is thicker so it has to be shorter* (spontaneous attempt to dissociate the weight from the volume).—So what about (Pb + Ib) and (Ia + Ic)?—*They're the same.*—And what about this ball of clay and the lead?—*That's easy.'* He weighs them. *'Oh, they're the same.*—And the clay and the blue bar?—*They'll be the same as well. No, the clay is lighter, we saw that it was with the lead. Oh no, they're the same.*—And what about (clay + lead) against (red + blue)?—*They're the same. If it was a little thinner* (= less dense) *the lead would be as* (long as) *the bar just as I said before.*—And what about these (red + blue) against these (clay + red)?—*We saw that they're the same.*—And what about these (clay + blue bar) against those (lead + red)?—*They're still the same.'*

FRED (7; 6) examines the bars and straightaway says that A = C if A = B and B = C. 'What about the blue and the lead?—*The bar will be heavier because it's longer.* (Weighing.) *Oh no, they're the same.*—And the black and the lead?—*They're the same as well, because the black and the blue are the same length together.*—And against the red?—*It's the same with all of them.*—What about these (red + the lead) against these (blue + the black)?—*They're the same. No, the blue and the black are heavier.*—Why?—*Because there are two of them.*—Do you remember the red and the lead?—*Yes, they were the same weight.*—And (the red + the lead) against (the black + the blue)?—*They're not the same weight; oh yes, the bars are the same as the lead.*—And what about this clay and the blue bar?—*The clay will be heavier.* (Weighing.) *They're the same.*—And this (the blue + the black) against that (the red + the clay)?—*They're the same because the red weighs as much as the clay, and also as much as the other bars.*—And these (the red + the clay) against those (the black + the blue)?—*We've got two bars and the lead and the clay. If I held the two bars in my hand, the black would be like the lead and the clay like the blue, so I think they must be the same.'*

TEL (7; 6). 'The blue and the red?—(Weighing.) *They're the same.*—And the red and the black?—*Also.*—And the blue and the black?—*The black looks a little bigger; no, it's the same because they all weigh the same.*—And the lead with the blue?—*The lead will weigh more because it's thicker.* (Weighing.) *No, it's the same because this one* (the red bar) *is long but that one* (the lead) *is thick.*—And with these (the red, etc.)?—*It's the same, because the blue and the lead are the same weight, and the black and the red are the same as well.*—And what about this (the red + the blue) against that (the lead + the black)?—*Oh, that's the same again, because all the bars are the same size. No, I'm wrong, the bars are heavier.*—And the (blue + the lead) against the (red + the black)?—*They're not the same. The lead is thicker and this one* (the blue bar) *is longer, and on the other side we have two long ones, so the bars must weigh more.*—Do you remember weighing the lead and the blue?—*Oh yes, they're all the same.'*

Clay and bars: immediate addition of the equivalences.

CAS (7; 11). 'The lead and the blue?—*The lead is heavier.* (Weighing.) *No, it's the same.*—And with the red?—*It's the same as well. It doesn't matter which because the three bars are the same* (checks with the scales).—And the (lead + the black) against the (red + the blue)?—*The lead side is bound to be heavier, because the lead is heavier than the bar.*—But when you weighed it before, how heavy was the lead?—*The same as a bar, but I'm sure it's heavier all the same.'* He weighs the lead and a bar once again, and falls silent, scratching his head in clear embarrassment. *'They're the same.*— So what about all these (the four objects, taken two at a time)?—*They're the same. There'll still be four with the bar and the lead; they're the same weight, so when you add a bar to the lead they ought to be the same as well.*— Why?—*Because the lead weighs as much as a bar, so two bars must be the same weight as one bar and the lead.'*

Having established that the lead has the same weight as the clay ball, Cas is asked: 'Are these two (lead + clay) the same weight as those (two bars)?—*They ought to be the same.'* Carbon and lead: *'They're the same weight, because the carbon is light and big and the lead is heavy and small.'* But when the demonstrator puts three bars in one pan and the carbon + the clay + the lead in the other, Cas at first refuses to grant that they will balance. 'Are they the same weight?—*No, the carbon is lighter than the clay and the lead, but it's not lighter than any of the bars.*—So?—*So this lot* (the three heterogeneous objects) *must be lighter, because the bar is heavier than the carbon ... Ah no, it's the same weight; you can see it is.'*

QUIS (8; 4). 'What about the red and the black bars?—*They're the same.*— And the blue and the black?—*The same.*—And the red and the blue?— *They're the same weight, because the red and the black weigh the same and the blue and the black as well.*—What about the blue and this piece of clay?— *The clay is heavier.'* He weighs it and laughs. *'I was wrong.*—The red and the clay?—*They're the same, because the red is the same as the blue.*—The black and the clay?—*Also.*—And these (Ia + Ib) against those (Ic + clay)?— *The two bars are heavier.*—Why?—*On this side, there's only the clay and that bit of iron, but on the other side there are those two bits.*

'What about the lead and this bar (Ia)?—(Weighing.) *They're the same weight.*—And these (Pb + Ia) against those (Ib and Ic)?—*The two bars are heavier.*—Why?—*As we saw before, there are two bars on one side and only one bar and a bit of lead on the other.'* The demonstrator removes a bar from each side. 'And now?—*Now they weigh the same. It's like with the blue bar and the lead.'* The bars are put back. 'And like that?—*Ah, they're the same.*

'And these (clay + lead) against those (two bars)?—*These* (the bars) *are heavier.*—What weighs as much as the lead?—*A bar.*—And as much as the clay?—*A bar as well.*—So?—*Ah, the two pans weigh the same.'*

These illustrations of the gradual discovery of the composition and quantification of weights are of great importance not only to this particular analysis but also to the psychological study of quantities in general.

The reader will remember that subjects at Stage II proved quite incapable of deducing $A = C$ from $A = B$ and $B = C$, when the compositions bore on objects of different densities and hence on the weights as such, but that they had no such difficulties with objects of the same density. However, at Stage III they readily apply the same schema to two homogeneous bars and one object of different density (lead, clay, carbon, etc.). Thus Mur, having seen that the lead had the same weight as the blue bar, concluded straight-away: 'It's the same as all of them, because all the bars weigh the same.' Bod, too, realized immediately that the weight equivalence of the lead to the bar extended to all the rest, and so did Fred, Tel, Cas and Quis. Cha was the only one to check the equivalence with two successive bars before generalizing his conclusion, and this is precisely why we have treated him as an intermediate case (between Stage II and Stage III).

How are we to explain this relatively sudden advance, this sudden release of deduction? If we compare the responses of these subjects to those set out in Chapters 1–6, the answer is quite clear: at Stage II, where only the quantity of matter is conserved and the weight has not yet become a constant, formal reasoning based on the simple co-ordination of equivalences can only be applied to the quantity of matter (homogeneous composition) but not yet to the weight as such (heterogeneous composition). However, as soon as the weight itself is treated as a constant (Stages IIIA and IIIB), the child can begin to construct logical arguments of the type $(A = B) + (B = C) = (A = C)$. In other words, while weight remains a subjective quality it does not lend itself to any kind of deduction, however elementary. Thus, even when the child sees for himself that the lead balances the red bar and that the red balances the blue, he will not admit that the lead must also balance the blue: since the lead has quite unexpectedly turned out to be no heavier than the red bar, he concludes that it is bound to catch up with the blue. But, once he abandons this egocentric and phenomenalistic approach based on tactile impressions and realizes that weight is a physical constant, there is nothing to prevent him from grasping that if the lead weighs as much as one of the bars, it will also weigh as much as the others.

The logical operation leading to the co-ordination of equivalences is therefore of a piece with the physical operation leading to the construction of physical constants, and we must now look more closely at this matter.

217

When we do so, we find that obstacles in the path of additive compositions are as instructive as the successful co-ordination of simple equivalences, because they help us to follow the progress of logical grouping and hence of quantification step by step.

Let us note, first of all, that some subjects at Sub-stage IIIA are less successful than others in coping with four objects placed in two sets. Thus, Cha, our intermediate case, failed with them completely, while the rest solved the problem more or less quickly. But all of them showed an initial resistance to the additive composition $(A + B) = (C + D)$, and this resistance was not broken by a sudden flash of insight but only disappeared after several trials and errors.

There were, first of all, those subjects who could not go on from the simple to the additive composition of heterogeneous as distinct from homogeneous equivalences. In other words, they realized that the weight of the lead is equal to that of every bar when it is weighed against it directly, but they thought that the equivalence ceased as soon as an extra bar was added to each pan. Thus, Cha realized that the lead is 'the same' as every one of the bars taken in turn, but refused to grant that the lead plus one bar was equal in weight to two bars, on the grounds that 'the lead is heavier'. And when this contradiction was pointed out to him, he justified his answer by saying 'here there are two of them and over there it's just one'; thus completely ignoring the lead. Similarly, Mur said that the lead was 'the same as all of them (the bars), because all the bars weigh the same', but went on to argue that if the lead is added to one of the bars the combination is no longer equal in weight to two bars 'because the lead is thicker'.

Next, there were those who kept vacillating until they finally reached the correct solution.

Bod's case is particularly interesting, because it illustrates the construction of unity. He quickly saw that if the lead was equal in weight to one of the bars, it must also be equal to the rest, but nevertheless contended that if it were combined with the black bar, it 'will be heavier because the lead has more weight'. He then recalled the equivalence but nevertheless refused to disregard the inherent heaviness of the lead: 'The two bars will still be lighter.' Finally, reminded that the lead had the same weight as a bar, he decided to convert the lead into a unit like the rest, and this by means of an argument that enabled him to dismiss the special quality of this metal: 'If it was a little thinner, the lead would be as long as the bar'; whence he concluded that 'they are the same'. Fred, for his part, having declared that the lead 'is the same' as any of the bars, went on to say that two bars were heavier than one bar plus the lead, 'because there are two of them', thus placing equality before numerical unity. But, as soon as he was reminded of the equivalences, he proceeded straight to numerical quantification, and even offered an illustration of why the lead and the clay could be substituted for two bars: 'If I held the two bars in my hand, the black would be like the lead and the clay like the blue, so I think they must be the same.'

Quis arrived at the same solution, though only after some help from the demonstrator. At first he claimed that the two bars must be heavier than the clay plus one bar, because 'on this side there's only the clay and that bit of iron, but on the other side there are those two bits'.

Finally, there were those subjects who offered alternative solutions, and also explained their reasons for them. Thus, Cas, who granted that the lead was equivalent to any of the bars ('it doesn't matter which because the three bars are the same'), was, nevertheless, reluctant to conclude that the lead and the black bar must be equivalent to the red and the blue. His curious explanation went a long way to demonstrating how difficult these subjects must find the construction of sets: the lead is admittedly equivalent to a bar, but only in isolation: when it is combined with a bar, 'the lead is bound to be heavier'. This distinction between the lead 'by itself' and the lead in combination with a bar proves better than anything else how resistant intuitive qualities are to operational composition. However, after repeating his original weighing, Cas changed his mind and enunciated his principle of quantification in the clearest possible way: 'The lead weighs as much as a bar, so two bars must be the same weight as one bar and the lead.' With six objects he hesitated once again, but finally came down in favour of the composition: 'It's the same weight; you can see it is.'

In short, children at Stage IIIA are torn between two conflicting attitudes. On the one hand they cling to the subjective and egocentric evaluation of weights they have inherited from earlier stages: the lead must be heavier than the brass bar because no matter what the balance says the bar feels or looks lighter. On the other hand, they use a logico-arithmetic approach, thanks to which they can compose the weight into equivalences or differences, recompose it in accordance with the logic of classes or relations, and quantify it by the combination of the two. When the composition stops short at term-by-term equivalences $(A = A') + (A' = A'') = (A = A'')$ or term-by-term differences $(A \leftarrow B) + (B \leftarrow C) = (A \leftarrow C)$, the second approach tends to overshadow the first; but when it goes beyond these elementary limits, and calls for greater co-ordination, then the simpler approach takes over. This is why one and the same subject will contend that the lead is equal to the bar when they are compared singly, but not when they are compared in pairs or larger sets. The result is a latent conflict between the false logic of direct experience or egocentrism and reversible composition, and it is only after a series of trials and errors that the latter carries the day.

Stage IIIA reactions therefore pose the following two problems: why do these children find it easier to deal with certain simple equivalences than with others or with additive compositions, and how are the latter constructed both logically and numerically?

The first of these two problems is easy to solve, the more so as it is identical with the one we encountered in connection with Sub-stage IIIA responses to the composition of inequalities (Chapter 10). There we saw that these children are capable of co-ordinating $A \leftarrow B$ and $B \leftarrow C$ into $A \leftarrow C$, but that they still rely on trial and error when it comes to the seriation of four terms $(A \leftarrow B \leftarrow C \leftarrow D)$ taken two at a time. Now, since the seriation of four terms calls for the same operation as that of three terms, the only possible explanation of the difference is that the second is intuitively simpler. In other words, at Sub-stage IIIA seriation is not yet fully operational. Now, the situation is precisely the same in the present case. Before he can conclude from $Pb = Ia$ and $Ia = Ic$, etc., that $Pb = Ib = Ic$, etc., the child must discard phenomenalism in favour of an operational system. But it is equally clear that, in the present case, the construction of that system is greatly helped by the intuitive equivalence of the bars. Thus, as soon as we introduce the lead, the clay and the carbon, the system tends to break down. In the case of additive compositions, even as simple as $(Ia = Pb) = (Ib + Ic)$, we have the further problem – and this is of great importance – that the terms admitted to be equivalent when they are compared one by one, must now be compared two at a time. Consequently the intuitive factor which facilitated the substitution of Ib or Ic, etc., for Ia in the equivalence $Ia = Pb$, becomes an impediment: this time a homogeneous set $(Ib + Ic)$ must be compared with a heterogeneous set $(Pb + Ia)$ and the very fact that the two are not deemed equivalent proves that the simple equivalences constructed at this sub-stage are still partly based on intuition. In other words the child, having granted that the weight of the piece of lead may be equivalent to that of a bar of brass, by temporarily disregarding its other qualities, goes on to grasp that it must also be equivalent to all the other bars whose identity he supposes intuitively, but once the lead is introduced in a new set, the equivalences disappear and the intuitive qualities resume their previous role. As a result, the lead once again becomes heavier in itself.

This semi-intuitive and semi-operational approach by subjects at Sub-stage IIIA throws a good deal of light on the logical and numerical construction of additive compositions, and makes it clear how the latter gives rise to intensive as well as extensive quantifications.

Logical addition, first of all, is nothing other than a combination of the elements (for example, $Pb = A_1$ and $Ia = A'_1$) into the class $A_1 + A'_1$ (or $Ib = A_2$ and $Ic = A'_2$ whence $A_2 + A'_2 = B_2$) or of two classes into a general class, for example $B_1 + B_2 = D$. Now, this combination can be effected intuitively as soon as the child realizes that the elements remain what they were no matter what their re-arrangement, in other words as soon as he grasps that the composition is reversible, and hence that the whole and its parts are conserved. The reader may recall that the non-conservation of weight (Chapters 2 and 5) was due precisely to the non-fulfilment of these conditions. Now, when our present subjects believe that the weight of the lead (A_1) changes, depending on whether they compare it to a single bar or incorporate it in the sets B_1 or D, they argue in exactly the same way. But, in that case, how do they arrive at the additive composition? Once again, by the gradual co-ordination of the direct operation $A_1 + A'_1 = B_1$ with its inverse, $B_1 - A'_1 = A_1$, and with the identity operation $A_1 + 0 = A_1$, i.e. by a reversible construction that permits them to pass from one arrangement to the next without denying their earlier findings. 'It's the same', Cas and Quis said in the end, because 'we saw it before'. What we have here, therefore, is method 1 described at the end of Chapter 1.

But how does the child go on from the addition of classes to numerical compositions, i.e. to metric quantities? First of all by a generalization of all the possible substitutions (equivalences) within the classes B_1, B_2 or D. 'Because the lead weighs as much as a bar,' said Cas, 'two bars must be the same weight as one bar and the lead.' In other words, it is by disregarding the qualities of A_1 and A'_1 or of A_2 and A'_2 that these children obtain $A_1 = A'_1 = A_2 = A'_2$ whence $A_1 + A'_1 = A_2 + A'_2$; i.e. $B_1 = B_2$.

However, this unity-engendering generalization involves a complete re-organization of the operational grouping of classes and relations. Thus, if $A_1 = A'_1 = A_2 = A'_2$, the class B may be composed of $A_1 + A_2$, of $A_1 + A'_1$, or of $A'_1 + A'_2$, etc., and hence ceases to be a qualified class. Now this is precisely what the child refuses to grant at the beginning: to him the two bars $A_2 + A'_2$ make 'two' bits while the lead and the bar make 'just one' (Cha), i.e. the lead A_1 is not considered a unit like the rest.

Number, by contrast, appears just as soon as the child realizes that any pair whatsoever is equivalent to any other, and that any triplet whatsoever is equivalent to any other. Now this implies that the general correspondence between any one A and any other A, between any one pair B and any other pair B, etc. goes hand in hand with an equally generalized seriation; if $\xrightarrow{a}$ designates the first rank (= the difference in rank between the first A and 0); $\xrightarrow{a'}$ the difference in rank between the next highest A and 1A; $\xrightarrow{b'}$ the difference in rank between the next highest A and 2A, etc., we have $\xrightarrow{b} = \xrightarrow{2a}$; $\xrightarrow{c} = \xrightarrow{3a}$; etc. These ranks remain unchanged if the terms are inverted. If number and measurement thus constitute a system of units based on generalized substitutions, they do so thanks to the fusion into a single operational whole of the addition of equivalences, characteristic of the grouping of classes, with the seriation of differences in rank, characteristic of the grouping of transitive asymmetrical relations.

Let us now look at the completion of this logical-*cum*-numerical construction at Sub-stage IIIB, when all the necessary compositions, even those bearing on the weight of objects of different densities (heterogeneous compositions), are effected without hesitation. We shall begin with a curious transitional case:

DEP (7; 10) establishes that the red bar is equivalent to the blue, and the blue to the black. 'And will the red be the same weight as the black?—*You might say it's thicker and heavier.*—Than what?—*Than the red.*—But what was the red like?—*Like the blue.*—And the black?—*Also like the blue.*—So?' He weighs the black against the red. *'They are the same, I can see that they are.*—And the blue and the red?—*Oh, they ought to be the same thing as well; the red is like the black, and the blue is the same as the red.*

'What about the red and this piece of lead?—*The lead is heavier.*' He weighs it. '*No, I was wrong.*—Can you see something else on this table that will weigh as much as the lead?—*Yes, the black one, because it's like the red; and the blue one as well because it weighs as much as the black and the red.* —What about these (red and black) and those (blue plus lead)?— (Reflects.). *I can't really tell. They won't be the same because there are two bars. Oh, I remember, the lead weighs as much as the red, so the lead with the blue makes two, and two and two that's the same.*—And what about these (black plus lead) and these (blue plus red)?—*You're just changing them around, it's not worth the trouble, they're still the same weight.*

'And this (clay) with the blue?—(Weighing.) *It is still the same as with the red and the black!*—And the clay and the lead?—*It's the same, because the lead is like the red and the clay is like the blue. The blue is like the red, so the clay must be the same as the lead.*—And these (clay and lead) with those (black and blue)?—*They're the same.*—And if the carbon is like the red, what will the lead and the carbon make together?—*The same as two bars, like the blue and the black.*—Are you sure?—*I think so; yes, I'm sure, we haven't weighed the carbon with the black, but it goes with the lead and the lead went with the bars when we weighed them.*

'Can you arrange something that weighs as much as the three bars.' He puts down the clay plus the lead plus the carbon. *'We've weighed the clay with the black and saw they were the same. The lead is the same as the blue; and the carbon is the same as the red, so these three bars and these three things must weigh as much as each other.'*

GER (9; 6). 'Does this blue bar weigh as much as the white?—(Weighing.) *Yes.*—And does this red bar weigh as much as the blue?—(Weighing.) *Yes.*— And the red and the white?—*They ought to weigh the same because they went together.*—And the lead?—*The lead will be heavier.* (Weighing.) *Oh, no!*—And the lead and the red?—*They will be the same, because the red and the blue are the same.*

'What about these (white plus red) and these (blue plus lead)?—*They're the same because the white is like the blue, and the red and the lead as well.*— And what about this clay with the lead?—*It's lighter.* (Weighing.) *Oh, no; they're the same.*—And this (clay plus lead) with that (blue plus red)?— *They aren't the same. Oh yes, they are. No; two bars, that makes two and this one* (he has picked up the clay) *only makes one. But wait a minute!'* He substitutes the clay for a bar, thus being left with the lead and one bar in one pan and the clay and another bar in the second. *'Like this they are the same.'* The demonstrator restores the original arrangement. *'It's still the same.*—How do you know that the bar weighs as much as the clay?—*We've weighed the clay with the lead, and the lead weighs the same as a bar.*

'What about this (carbon) and that (clay)?—(Weighing.) *They're the same.*—Look, I've put three bars on this pan; can you balance them?' He puts down the clay plus the carbon and the red. 'Any other way?' He makes all the substitutions and says: *'It's always the same.'*

TIT (9; 6) establishes all the simple equivalences. 'What about these (blue plus white) and these (lead plus red)?—*They'll balance. The lead weighs as much as the blue, and the red and the white as well, so everything will be even.*— And if we change them about like this (the lead with the white)?—*They'll still be the same.*—Check if this clay weighs as much as the lead, will you?— (Weighing.) *Yes. It's because there is more clay than lead, so they are the same weight.*—And these (two bars) with those (clay plus lead)?—*They'll weigh the same because the clay weighs as much as the lead and the bar.*—And if we change them about?—*It's still the same; this bit of iron is the same as that other bit of iron, and the lead is the same as the clay.*—And the carbon?— (Weighing.) *Same as the lead.*—And these (clay and two bars) with those (one bar plus lead plus carbon)?—*They balance; the lead is the same as this white bit, and the red is the same as the blue, and the carbon is like the clay.*—Can you arrange them differently?' He makes all the substitutions. *'It's always the same.'*

LAR (10; 0) also grants all the combinations. *'Each one* (of the heterogeneous objects) *can take the place of a bar.'*

Such, then, are the final reactions of our subjects. Dep still showed some of the hesitations characteristic of Sub-stage IIIA when he wavered between subjective evaluation and objective composition, but as soon as he realised that the equivalence was transitive ('Oh, they ought to be the same thing as well'), he at once extended the additive composition to four or even to six objects, thus treating each as a numerical unit: 'The lead with the blue makes two, and two and two that's the same.'

But why is it that these subjects have no difficulty with deductive compositions, when those at Stage IIIA only arrive at them after a series of trials and errors?

The first answer to spring to mind is the intervention of a sudden structuring process, the sudden emergence of a *Gestalt*. Thus when subjects at Stage IIIA exclaim, after many hesitations, 'No, they are the same weight, you can see that they are', or when the present subjects declare straightaway, 'they ought to weigh the same because they went together' (Ger), or 'they'll balance' (Tit), we have the impression that there has been the kind of crystallization we normally associate with sudden changes in perceptive structures. However, if as Wertheimer suggests, the assimilation associated with deduction is indeed a process of formalization then it is clear that, in our particular sphere, the re-structuring must be the result of real operations, i.e. of combinations (logical or arithmetical additions and subtractions) and of substitutions, either singly or within the sets formed by the additions. How, in fact, does a child discover that $A = A''$ if $A = A'$ and $A' = A''$? Precisely by the mental substitution of A'' for A' in the equality $A = A'$. Now, that substitution, the basis of equivalence, is a real operation performed mentally or in action. Thus Ger, perplexed by the comparison of the two bars with the lead and the clay, suspended judgment until he had changed the clay for one of the bars to facilitate his calculation. Similarly Lar completed his mental substitutions with a series of physical substitutions before concluding that 'each one can take the place of a bar'. Unlike perceptive re-structuring, which is the discovery of the 'good form', logical structuring is 'formal' in quite a different sense: the logico-arithmetical 'good form' is one of groupings or of groups in which all the operations are reversible, composable and associative.

Now, the reactions we have just described involve no less than three groupings: the preliminary grouping of equivalences (or substitutions) $(A = A') + (A' = A'') = (A = A'')$; the additive grouping of classes allowing of the construction of any class whatsoever by conjunction $(+)$ or exclusion $(-)$, e.g. $A_1 + A'_1 = B_1$ or $A_2 + A'_2 = B_2$, etc.; and finally the additive group of integral numbers which makes it possible to treat every element as equal to, but distinct from, any others. How do our subjects pass on from the additive grouping of classes to the group of numbers? Once again, by generalizing a substitution or by introducing a general equivalence based on the exclusion of all qualities other than weight. As Ger said after his generalized substitution: 'It's always the same.' To him, equivalent units were no longer distinct except for the order of their correlation or enumeration.

§4. *Conclusion: physical and logico-arithmetical operations*

We have come to the end of our study of the groupings of serial operations and equivalences in the case of weight, and we have previously encountered the same logical compositions and the quantifications resulting from their operational synthesis when treating of the quantity of matter. The moment has therefore come to analyse the links between these logico-arithmetical operations and the physical operations at work in the elaboration of the conservation of substance, weight and volume.

Let us first of all recall some definitions. Logico-arithmetical operations are operations with classes (or conjunctions of equivalent terms), with asymmetrical relations (or series) and with numbers, while physical operations consist of partitions, displacements and measurements based on congruency. Now it is clear that the three elements of both sets of operations are in one-to-one correspondence: in the first set, space and time are simply disregarded and replaced by deductive exteriority and successions while, in the second set, they are not. Let us say straightaway that this distinction in no way implies that the first set of operations is effected mentally and the second materially: a class can be engendered by the construction of a 'heap' and a series by the alignment of objects in a 'row'; and in both cases we can disregard the space occupied and the temporal order of the successive operations. Conversely, we can mentally divide a lump of sugar into 'grains' and immerse them in a glass of water, only remembering that the state of dissolution is necessarily subsequent to the initial solid state. The characteristic shared by both types of operation is that they are reversible by definition; and the difference between them is not that the second consists of material actions and external transformations, while the first does not, but is purely spatiotemporal. For the rest, as we have said, these two sets of operations are in one-to-one correspondence. A logical class and its division into sub-classes or component elements is, in fact, a system of (additive or multiplicative) conjunctions and separations (subtraction or logical divisions), and the physical operations at work in the division or reconstruction of the initial whole constitute precisely the same system, but in the spatio-temporal field. Again, the seriation of asymmetrical relations is simply a juxtaposition of the differences, each of which thus constitutes a segment of the total series. The equivalent physical operation is a system of 'placements' and displacements. Finally, it is clear that numbers correspond to the units chosen for purposes of measurement, and that the physical counterparts of their composition are substitutions based on spatio-temporal congruencies. Divisions and displacements can therefore be treated both qualitatively or logically, in which case a body becomes a set of qualitatively identifiable fragments, and also quantitatively, thanks to measurements which, in turn, constitute an operational synthesis of the divisions and displacements, much as number is a fusion of a class with an asymmetrical relation.

For all that, logico-arithmetical operations cannot be considered the form of which physical operations constitute the content, because before the child can grasp that the whole remains constant no matter into how many parts it is divided (this grasp is, in fact, what distinguishes operational division from whittling), he must have learned to handle the (formal) logic of classes or numbers. The latter, we believe, helps him not only to construct the most general forms of composition but also those that can be applied to all others, much as a dress can be applied to the female body. Let us also note that the form and content of thought are purely relative concepts, because the physical operations that are assumed to constitute the content of the logical forms can, in their turn, be considered so many forms with an experimental content. Thus division is the form whose content is, say, the perceptible ball of clay; and the invariance of the weight of the divided whole is another form whose content is the result of experimental verifications, i.e. of weighings.

The relationship between form and content thus faces us with two problems: that of the links between logico-arithmetical and physical operations, and that of the links between the latter (or both) and experience. We shall study the second of these problems in the next chapter, but before we can do so, we must first try to resolve the first.

As we saw at the end of Chapter 9, the problem of the relationship between logico-arithmetical and physical operations has three possible solutions. In the first, the construction of the form is said to precede and determine that of the content, in which case logico-arithmetical operations must be elaborated before physical operations, so that the latter become the consequences of the application of logic to reality. It is only if we adopt this solution that we can speak of form and content in the classical sense. In the second solution, by contrast, the construction of the content may determine that of the form, in which case physical operations precede logico-arithmetical operations, and both spring from experimental induction, or more precisely from inductive experiment. In the third solution neither operation precedes or determines the other; both are constructed simultaneously, in which case experimental induction must be identified with nascent physical composition (or with logico-arithmetical composition in the case of inductions or analogies bearing on classes, series or numbers).

The first solution looks the most plausible at first sight. It is hard to deny, *a priori*, that the mind is incapable of applying formal logic to the concepts of weight and volume even before it discovers by physical operations that the ball of clay conserves its weight when its shape is distorted, or that the water level remains constant once the sugar has fully dissolved. It would rather seem that the child must have learned to seriate weights or to conclude that the weight A = the weight A″ if A = A′ and A′ = A″ before he can construct the invariance of the weight of a transformed ball or of a lump of sugar that apparently turns into clear water. However, as we saw in Chapter 10 and again in the present chapter, this interpretation is not borne out by psychological studies; neither Aristotle's logic of classes nor Russell's logic of relations can be said to have paved the way for the work of Galileo or Lavoisier, because it is impossible to construct either a logic or an arithmetic of weight before its physical aspects have been grasped.

What this chapter and the previous one have shown quite generally is that the same formal or logico-arithmetical compositions are not applied synchronously to matter, weight and volume, but that like the physical operations examined in Chapters 1–9, they are subject to a systematic time-lag. Quite evidently, therefore, the logico-arithmetical construction of each of the three invariants goes hand in hand with its physical construction. Moreover, our study of seriation (Chapter 10) has thrown a great deal of light on the reason for these time-lags: matter, weight and volume make distinct impressions on the child, both in action and in perception; so much so, that his conception of volume and weight remains egocentric and phenomenalistic much longer than that of the quantity of matter. Again, the study of the composition of weight equivalences (Chapter 11) has shown us that the absence of physical invariants always goes hand in hand with the absence of logical invariants, or of what Arnold Reymond[1] called 'functional invariants': until such time as he comes to treat the weight of an object as a physical constant, the child cannot impose weight equivalences either logically or arithmetically or even combine equivalent terms into sets of two or three, thus conserving the equivalences, no longer as physical constants but — and this is what is so astonishing — as simple logical constants, i.e. as data or premises

[1] A. Reymond, *Les Principes de la logique et la critique contemporaine*, Paris, Boivin.

that do not change in the course of an experiment. Thus $A = A'$, but if $A' = A''$ then A is not equal to A''; or again $A = A'$ but if A is combined with A_2 and A' with A'_2 then A is no longer equal to A'! In the next chapter we shall see that the same remarks apply to volumes; their own composition lags one stage behind that of weight, much as the composition of the latter lags one stage behind that of substance.

In short, the first solution, i.e. the assumption that logico-arithmetical compositions are elaborated in advance of the corresponding physical operations and that they determine them as the form determines its contents, must be rejected: the former entail the latter, and *vice versa*. We can only argue about weights once we have constructed their logical invariance with the help of non-contradictory concepts or relations; similarly the physical composition of weights implies the existence of physical invariants. Now the two types of invariant are interdependent and cannot be constructed independently of each other; the first, as it were, represents our axiomatic knowledge, and the second our real knowledge, and before the formal mechanisms have reached a high degree of generality it is exceedingly difficult to construct an axiomatic system in advance of the relevant science.[1]

The second solution which reduces logico-arithmetical to physical constructions and both to experience is, as we shall see in Chapter 12, equally unacceptable, because not only experimental induction but also the very reading of experimental data calls for compositions; true experience, unlike direct perception, is always a construction.

Hence the only valid solution is the third: that logico-arithmetical and physical operations develop hand in hand. How can we explain their synchronism? Is it the result of interactions or of separate but parallel developments?

The reader will recall that in the last chapter devoted to the seriation of weights, we concluded that the gradual grouping of asymmetrical or serial relations is due to the gradual elimination of logical egocentrism, according to which weight is an absolute quality depending on the child's own actions. More generally, all logical groupings of a system of concepts or relations call for the abandonment of egocentrism, and that is why the construction of one system can lag behind that of another more intimately linked to perception or direct actions. This also explains why the various logical compositions are not completed at the same time, but not why they are synchronous with the corresponding physical compositions, e.g. with those that go into the construction of our

[1] *Cf.* Gonseth, *Les Fondements des mathématiques*, Paris, Blanchard, 1926.

three invariants, of atomism and of the schema of compression or decompression. To find that explanation, however, we need only re-examine the arguments by which our subjects tried to justify their belief in non-conservation (Chapters 1–9). If we do so, we shall discover that the main obstacles to physical composition and to the construction of physical invariants are the perceptive appearances, i.e. phenomenalism or qualitative reality before it is corrected by reason, or rather before it is completed by a construction designed to 'save the appearances' ($\Sigma\omega\zeta\varepsilon\iota\nu$ $\tau\grave{\alpha}$ $\phi\alpha\iota\nu\acute{o}\mu\varepsilon\nu\alpha$!) by fitting them into the universe from which they are derived. Hence, much as the subjective view never completely disappears from logical compositions but is adjusted until it becomes one relation among all the other possible ones, and this precisely thanks to the decentration of the self by means of groupings, so the apparent relations do not disappear completely from physical compositions but are co-ordinated with the other relations thanks to the correction of phenomenalism in the light of rational experience. Nor is that all: phenomenalism which stands in the way of physical compositions, and egocentrism which retards logical compositions, are, as we have tried to show throughout this book, two aspects of a single illusion. Phenomenalism appeals only to those who have not transcended their own perceptive viewpoint, and egocentrism means mistaking one's own perceptions for the only possible reality. When we say that the moon seems to follow us about, we have a relationship that becomes objective as soon as it is incorporated into a group of displacements and perspectives; but when we say the moon follows us about 'in fact', we are expressing a phenomenalist view based on egocentrism, or an egocentric view based on phenomenalism. Similarly, the child who believes that a ball of clay becomes lighter when it is twisted into a coil because it has been pulled out of shape, because it is shorter, because it's thinner, etc., is combining logical egocentrism with physical phenomenalism. Hence, we need not be surprised that the development of physical operations should be synchronous with that of the associated logico-arithmetical operations.

This brings us back to our earlier question: are these two types of operation simply parallel or do they interact? Must the child construct physical invariants before he can reason about the corresponding concepts, or must he construct the logical invariants before he can construct the corresponding physical systems? When posed in this way the problem is purely artificial, because logico-arithmetical, i.e. non-temporal, relations occur in every physical construction, and spatio-temporal, i.e. physical, relations occur in every logical construction (combinations, seriations, etc.).

The real problem which we have encountered throughout this book, and particularly in Chapter 9, must rather be stated as follows:

An operation is a reversible action. In the case of logico-arithmetical operations, for example, it is clear that if we can combine two objects into a single set, e.g. $A + A' = B$, we can equally well retrieve one of them by subtraction, e.g. $B - A' = A$. Again if divisions and displacements constitute physical operations in contrast to random transformations, it is precisely because they are reversible even in the spatio-temporal sphere: a division can be offset by the re-combination (in fact, or in thought) of the fragments into the initial whole, and a displacement by a return to the original position. Now, as we have shown at length, it is precisely when he discovers the reversibility of these transformations that the child recognizes their operational character and becomes capable of explicative compositions. That is why, for example, he grasps that the dissolution of the sugar is a reversible process just as soon as he realizes that the lump splits up into grains which spread through the water and conserve their substance, weight and volume, but can be compressed back into a lump.

However, such operational reversibility presupposes the existence of elementary invariants and also leads back to them. From the physical point of view, a divided ball can conserve its original weight only if all the fragments conserve theirs, regardless of their position or displacement. Now, from a logical point of view, we can only effect a reversible composition as between the weight of a bar A and that of an object A′ into the whole B, if A has the same characteristics when it is combined with A′ as it has when it is not so combined. Now, we saw that the child questions the existence of these elementary invariants no less than that of the total invariant, and this for two reasons; his mistaken views of reality (due to real obstacles introduced by his phenomenalism), and his failure to handle reversible operations (due to formal obstacles introduced by his egocentrism). In the sphere of physical operations (see particularly Chapter 9) he finds it difficult to grant that the several fractions (half or quarter) of an object have one half or one quarter the weight of the whole, because the whole is thought to change its properties upon divisions and displacements (real obstacles); moreover, he is not sufficiently familiar with reversible compositions for his operational mechanism to triumph over his hesitations (formal obstacles). In the sphere of logical operations we saw, similarly, that even when he grants that $A_1 = A_2$, he is not sure that $A_1 + X = A_2 + X$ and this (1) because an object combined with another changes its intuitive appearance (real obstacle); and (2) because he again lacks the operational mechanism that alone would help him to shed his hesitations (formal obstacles).

How are these two obstacles eventually surmounted? Does the formal mechanism correct the child's mistaken ideas of reality, or is it the correction of these ideas that enables him to effect formal compositions? This is the real problem of the relationship between logical and physical operations. Now there is only one possible answer: the two obstacles are surmounted simultaneously. It is not the construction of invariants that ushers in the reversibility of operations, or *vice versa*: the two notions emerge together. Reversibility is constructed with the help of hypothetical invariants, and the very success of the resulting composition attests to their validity. Hence formal logic cannot possibly precede material compositions, and both must necessarily go hand in hand.

But in that case logico-arithmetical and physical operations must be identical from the outset, except for the fact that the former constitute a regulation of an operational mechanism, and the latter a regulation of the material or external results. It is only in the course of the child's further development that they become dissociated and differentiated, the first leading to a general logic and the second to the construction of the physical universe. But originally all operations are logical and physical at once. Before the age of six or seven years the child treats all numbers as mere figures, and all logical entities as complex objects whose collective aspects are classes and whose inner structure determines all their relations. It is only when he transcends this intuitive level and progresses to reversible operations that he begins to distinguish physical from logico-arithmetical operations. In their formal mechanisms both involve precisely the same transformations, but the first apply to the object and to its parts, or to its internal spatio-temporal links, and the second to collections of objects (classes), to links between objects conceived as elements of classes, or to links between classes (relations) or to both links at once (numbers). As this is the sole difference between the two types of operation, it is only to be expected that their qualitative or intensive groupings and their numerical or metric quantifications should be synchronous.

But in that case we have to solve a final problem. We have argued that physical operations constitute a composition of the external world, but does reality lend itself to such composition without further ado? There might, in fact, be incomplete compositions or results that can be reached empirically before they can be composed. What, in short, is the relationship between physical composition and what we commonly call experimental induction? These are some of the questions we shall be trying to answer in Chapter 12.

231

Simple and Additive Compositions of Equivalent Volumes and the Discovery of the Displacement Law

The final chapter of this book has two aims: to conclude our study of logico-arithmetical compositions, and to clarify the relationship between rational compositions and experience.

To that end we shall first try to correlate physical operations bearing (1) on the quantity of matter, (2) on weight, and (3) on physical volume with the associated logico-arithmetical operations. Now, since (1) has been examined at length in our earlier studies of seriation and of the composition of continuous and discontinuous quantities in general; and since (2) has been discussed in Chapters 10 and 11 of this book, we need only concern ourselves with the case of volumes. Accordingly, we shall, in the first part of this chapter, apply to the volume of solids immersed in water precisely the same composition we have applied to its weight, except that we shall not be devoting a special section to changes in volume, because the problems the child must solve to compose equivalent volumes tells us enough about how he would treat such changes. This discussion will, moreover, enable us to re-examine our earlier assumptions about the relationship between physical and formal or logico-arithmetical operations.

However, the question of physical volumes has yet another aspect on which we must dwell at some length, because it will lead to the solution of a problem we left unsolved in the last chapter: that of the relationship between rational operations and experience, i.e. between deductive compositions and experimental induction or the discovery of physical laws.

We shall be examining both aspects with the help of a familiar method, namely the immersion of various solids in a vessel filled with water to the three-quarters mark and the determination of their volumes by the rise in the water level. Now, this method tells us a great deal about the child's conception of the phenomena he observes: to grasp what is happening he must not only assume the conservation of the volume of the immersed solid despite changes in its position, etc., but also, and above all, the conservation of the volume of the displaced water. Now, the reader will recall that small children who have not yet differentiated the quantity of matter from the weight and the volume believe that it is the weight, not the volume, of the immersed solid which makes the water rise. Hence, when we were merely concerned with the conservation of the volume of the immersed solids (Chapters 3 and 6) we removed this obstacle by explaining to our subjects that the water rises because of the 'place' the solid occupies, and not because of its weight. But, now that we are concerned with the relationship between the composition of volumes and experimental induction, we must consider both aspects of the problem, i.e. the volume of the water displaced and that of the immersed solid. When we recall that children at Stages I–III base their belief in the non-conservation of physical volumes on the instability of matter, i.e. on the view that all deformations lead to changes in texture, their assertion that the water rises because of the weight of the immersed solid can only mean that they think the respective volumes of the water and the solid reflect a kind of struggle, the protagonists of which are the weights. The weight of the water, as it were, tries to flatten the pebble or ball, while the weight of the immersed object tends to push the water up. This is precisely why the position of the immersed object is thought to be so important: if it stands upright it will exert a greater or smaller force on the water than when it lies on its side, and its volume will depend on whether it wins out over the water or *vice versa*. The discovery of the correct law, i.e. that the rise in the water level is due solely to the volume of the immersed body, thus hinges on several complex factors: the dissociation of the weight from the volume, the conservation of the volume of the solid and of the liquid, and the physical-*cum*-logico-arithmetical composition of all the volumes involved. That is why we have reserved the discussion of the whole phenomenon for this final chapter, in which we shall be examining the composition of volumes and the inductive discovery of the law expressing the rise in level in terms of the relationship between the volume of the immersed solid and the volume of the displaced liquid. This law, as we shall see, reflects the victory of rational composition over egocentric pre-relations and over the phenomenalistic illusions of immediate experience.[1]

[1] We have also dealt with this law in *The Child's Conception of Physical Causality*, but in terms of the relationship between prediction and explication. In what follows, we shall be more concerned with the gradual composition of the logical operations involved, and thus develop a point of view we first advanced somewhat cautiously in *Judgment and Reasoning in the Child*, Routledge & Kegan Paul, 1928.

Our method will be parallel to that used in the earlier experiments: the study of simple equivalences and of additions of equivalent terms with objects of different shape or density. To enable the child to verify these equalities after he has predicted or denied them, we produce two identical measuring jars containing the same amount of water (roughly up to the three-quarters mark); these two vessels thus play the same role as the pans of the balance did in the last chapter. The child is asked for the following compositions: (1) To equate the volumes of objects of the same shape and weight, e.g. of three aluminium cylinders differing in colour only, (red = R, orange = O, and black = B). It is essential to ask him if these equivalences are independent of the position of the cylinders, e.g. whether a recumbent cylinder takes up the same space as a vertical cylinder, etc. (2) To equate objects of the same shape and volume but of different weights. To that end, he is asked to compare one of the aluminium cylinders to a brass or lead cylinder of the same shape and volume. (3) To equate the volumes of objects of different weight and shape, e.g. a coil of modelling clay, a ball of soft wax, etc. (4) To proceed to additive logical or numerical compositions of homogeneous units: to equate two aluminium cylinders of order I with a cylinder of double height (order II), or to equate three cylinders of order I with one cylinder of order III. (5) To proceed to additive, logical or numerical compositions of heterogeneous objects compared two or three at a time.

The responses of our subjects could be fitted into the four stages we have distinguished throughout this book, the first three corresponding to the three stages described in the last chapter (composition of weights). During Stage I there is no composition of any kind: the child is totally incapable of grasping the displacement law. At Stage II he is successful with compositions based on equal weights, but his compositions do not extend to objects of heterogeneous shape and are still based on inductive analogies rather than on logical necessity. He has still a marked resistance to additive compositions. This is also the stage when weight and volume begin to be dissociated, the first step in the elaboration of the displacement law. During Stage III the formal mechanism begins to emerge: compositions based on simple equivalences become rigorous and cease to be purely analogic; they can now be applied to objects of heterogeneous shape and the child also begins to succeed with additive compositions though, at first, he still tends to reserve his deductions for the weight factor; these compositions are therefore comparable to the Stage III compositions described in the last chapter, Weight and volume are only dissociated during the course of the tests, so that the displacement law is only discovered by steps, i.e. constructed by induction. During Stage IV the volume is completely dissociated from the weight, all the compositions are based on the volume, and the displacement law is elaborated by deduction.

§1. *Stage I: lack of composition and of experimental regularity*

Subjects at Stage I fail all the composition tests, even those involving three identical cylinders of different colour:

HAE (6; 12). 'What do you think will happen if I drop this (cylinder O) into the water?—*The water will come up to here.*' He points to a higher level. Experiment with Flask I:

'And if I drop this one (B) in here (Flask II)?—*The same.*—Why?— *Because that thing which goes down makes the water rise.*—Why?—*Because the weight goes to the bottom of the water and makes it come up.*—What if I drop the black one into this jar, and the red one into the other?—*The water of the red will rise higher because the red is heavier.*—Aren't they the same weight?—(Weighing.) *Oh yes. It's going to rise as high.*—And the red and the orange?—*The water will rise less with the orange.*—Why?— *Oh, well, because they are the same weight.*

'What if I put that one in like this (R horizontal), and the other one like that (O vertical)?—*The water will rise more with the orange, because it takes up more room.*—Why?—*It's bigger.*

'And this (the lead cylinder)?—*Oh, it's heavier than the other three.*— Will the water rise as much with the lead as with the red one?—*No, it'll go higher.*—Why?—*Because the lead is heavier.*—Look. (Experiment.)—*Oh, it's the same.*—And if I put in the lead and the black?—*The water will rise more with the lead.*—Why?—*Or, perhaps it's the same... No, it'll rise more.*— But you just said it might be the same?—*No, it'll go higher because I know the lead is heavier.*—(Experiment.) Why is it the same?—...—What if I put these (O + R) in the first jar and those (B + Pb) in the other?—*The water will rise with these (O + R) but it'll rise higher with those (B + Pb).*— Why?—*Because there are two of them, and the lead is heavier.*—And what about these (O + R)?—*They are two as well.*—Did you notice that Pb = B?— *Yes.*—So what about (Pb + B) and (R + O)?—*These (Pb + B) are going to make the water rise higher.*—Why?—*They're heavier.*—Just look! (Experiment.)—*Why are they the same?*—...—What about this (coil of clay = CL) and the red?—*The red will make the water rise higher, because it's made of iron.*—Look! (Experiment.)—*Same thing.*—And these two (CL and B)?— *Also the same... No, this one (B) will make the water rise higher because it's heavier.*—(Experiment.) And what about CL and Pb?—*The lead will make it go higher.*—We saw that the clay was the same as what?—*As the black.*—And the lead?—*Also the same as the black.*—So what about these (CL and Pb)?—*The lead will make the water rise higher because it's heavier.*'

EY (6; 5). 'What if I drop this one (O) in the water?—*The water will rise.*— And that one (B)?—*The same.*—And this (the red which he has just identified with the orange)?—*It'll rise higher.*—Why?—...—Which one did we just look at?—*The red and the orange.*—And before?—*The black and the orange.*—Look, I'm going to put this one in like that (B upright) and that one like this (O horizontal). What will happen?—*It won't rise the same.*—Why not?—*Because this one is standing up and it's thicker.*—Look. (Experiment.)—*Oh, they're the same.*

'Hold this (lead cylinder).—*It's thicker.*—What if we try these (Pb and R)?—*The water will rise higher with this one* (Pb) *because it's heavier.*—Look! (Experiment.)—*They are the same!*—What about these (Pb and B)?—*It'll be as I said before.*—Why?—*Because this one* (B) *is light, and that one* (Pb) *is heavier.*—And how high will the water rise?' He points to two different levels. (Experiment.) 'And what about these (Pb and O)?—*One will make the water rise more and the other less.*—Why?—*Because this one* (Pb) *is heavier.*—What happened with these (Pb and R)?—*They were the same* (correct recall).—And these (R and O)?—*Also the same.*—And these (Pb and O)?—*This one* (Pb) *will make the water rise higher.*—See for yourself. (Experiment.)—*Oh, it's the same.*'

Ey is shown that the lead displaces as much water as each of the cylinders O, B and R: 'Now look at these (Pb + B) in this flask and at those (R + O) in the other flask. Will they make the water go up as much as each other?—*The glass ought to get brim-full with these two* (Pb and B).—Why?—*The water will rise higher with* (Pb + B) *because they're heavier.*

'Look at this (CL). Will it make the water rise as much as the black one?—*No, the sausage is bigger.*—Let's try. (Experiment.)—*It's the same for both.*—That's odd, wouldn't you say?—*Yes.*—Well now, look. Do the sausage and the red one make the water rise as high as the sausage and the black one?—*No, it's more with the sausage.*—Why?—*It's thicker.*—And are the red and the black the same?—*Yes.*—And the sausage and the red?—*No, the water will rise more with the sausage.*—Look. (Experiment.)

'And if I put this (CL) in one glass and those two (R + O) in the other?—*The water will rise more with the sausage.*—Why?—*It's bigger* (= longer than the two cylinders).

'And how far will it rise with this one (cylinder of order III)?—*The glass will be full.*' (Experiment.)—'Mark it with an elastic band. And what about this (the red in the other vessel)?—*Up to here.*' (He marks a very low level.) 'And these (R + O)?'—(He marks a slightly higher level but quite at random.) —And these (R + O + B)?—*Up to there* (still higher, but well below the level of the big cylinder).'

BRA (6; 8) discovers that the three cylinders of order I produce the same rise in level: 'Why do they?—*Because these two* (he points to B and O) *are the same size, just look.*—And what if I put the black one in like this (vertically) and the orange one like that (horizontally)?—*When it stands up it'll make the water rise higher.*—Why?—...'

The three cylinders are immersed successively. For the first, he points to the level he has just observed; for the two he points to a lower level but corrects his mistake after an experiment; for all three he points to much too high a level. 'Now, look carefully (at the cylinder III, which is placed upright on the table by the side of the three cylinders superposed vertically).

You see? How far will the water rise if I put the big one in this glass and the three little ones in the other?' He points to much too low a level for the large one: *'Over here* (the three) *it's more because there are three of them, and over there it's just one.* (Experiment.) *Ah, it's the same because they're the same size, I mean the same weight.*—And now (the cylinders are removed and the measuring jars changed round) I'm going to put the big one in here and the three little ones together (still superposed) in here. What do you think will happen?—*Perhaps the long one will make the water rise more because it's long.*

'What about the red and this one (Pb)?—*The lead will make the water rise higher because it's heavier.* (Experiment.) *Oh no, it's the same.*—Do you remember the red and the black?—*Yes, they were the same, because they are the same size.*—So what about these (Pb and B)?—*They're not the same; the water will rise higher over here because this one* (Pb) *is heavier.* (Experiment.)

'Well, if they're the same, then I'll now put these two (R + O) in one glass and those (Pb + B) in the other. What will happen?—*The water will rise higher over here* (Pb + B), *because these two include one that is heavier.'*— (Experiment.) How do you explain that?—*Perhaps they're the same weight?*— What about these (Pb and B)?—*No, they aren't.*—So?—...'

MAG (6; 10) produces similar responses, so that we need merely report his reaction to the additive compositions. 'What about these (R + O + B superposed) in this jar, and that one (III) in the other?—*The water will rise higher over here, because this one* (III) *is longer.*—Look!'—(Experiment.) *'They make it rise the same because they are as long as each other* (which he has seen before the experiment!)—And like that (the three small cylinders piled up horizontally)?—*The water will rise more with the big one.*—And like that (the three small cylinders piled up vertically)?—*The same, we saw that it was.*—And like that (two small cylinders superposed vertically and the third placed horizontally on the top)?—*No, that won't be the same.'*

These responses are most revealing, but before we discuss them in detail we must first deal with a possible objection. The child comes to the test with two preconceived ideas, namely that the weight, the volume and the quantity of matter are more or less proportional to one another, and that the rise in level is caused by the weight: he believes that thicker bodies (as Ey said of both their volume and weight precisely because he thought these properties were identical) are also stronger and hence more capable of lifting the water. Now, instead of undeceiving him in so many words (as we did in Chapter 3, where we were looking for evaluations of the volume of the submerged substance), we simply present him with a series of volume correlations. But in that case might he not simply rely on 'affective' logic and, instead of looking for a new law as we expect him to do, merely try to justify his original answers, come what may? We do not believe that children do this, the less so as we make sure at the beginning of the test not to demand a hypothesis that they may later have to defend or reject. Their presuppositions remain implicit and simply constitute so many intellectual obstacles in the path of the displacement law, and since no one ever reasons unless he is faced with a difficulty (in its absence, direct perception or the memory take the place of reasoning), the situation is perfectly normal. In what follows we shall, however, distinguish between (1) logico-arithmetical compositions as such, and (2) specific reactions to the experiments or to the law relating level and volume.

In respect of (1) it is obvious that these children are at a loss with even the most elementary compositions. Take the case of simple equivalences. Hae, having established that the orange and black cylinders are equivalent, was not at all sure about the black and the red even though he knew that they were of equal dimensions; moreover, having established that $B = R$ he failed to conclude that $R = O$. Ey granted that the red and the black produced the same effects, and also the orange and the black, but not the red and the orange. Admittedly, had the three cylinders been left unpainted, all our subjects would have grasped that they were of equal volume (as they did with the brass bars in Chapter 11), not by virtue of the composition $(A = A') + (A' = A'') = (A = A'')$, but merely because there were no discernible differences between the cylinders.

Moreover, if we simply alter the position of one of the two elements whose identity has been admitted, our subjects think that it will no longer produce the same rise in level as the other. Thus, Hae said of one horizontal and one vertical cylinder: 'The water will rise more with the orange (i.e. the vertical cylinder)' because the orange 'takes up more room ... it's bigger'. Ey, too, thought that the vertical cylinder would produce a greater rise in level 'because this one is standing up and it's thicker'; and Bra said: 'When it stands up it'll make the water rise higher.'

In the last chapter we saw that children believe a brass bar changes its weight depending on whether it is placed parallel or perpendicular to another bar, but the illusion is much greater in the present series of tests, and it also persists much longer (until Stage III). The reason is quite simple: our subjects not only assume that the weight of the cylinder (like that of the bar) is transformed by changes in its position,[1] but that its size, and hence its volume, change as well, and this despite their normal belief that objects are of constant shape and dimensions (which is far from obvious when the objects are immersed in a liquid).

Finally, the child must consider not only the volume of the cylinder but also that of the displaced water, which he is inclined

[1] Moreover, children up to and including Stage II also believe the same of a disc of clay.

to think depends on the pressure exerted by the cylinder and hence on its position. The combination of these three factors suffices to explain why the illusion persists so much longer with the cylinders than with the bars.

Equivalences involving two homogeneous objects and one heterogeneous object prove to be equally incoherent, but on the purely logical plane. Thus when Hae discovered, contrary to his prediction, that the lead produces the same rise in level as the red cylinder, he concluded that the black cylinder would make the water rise higher. And yet he remembered perfectly well that $B = R$ because he added: 'Or perhaps it's the same ... no, it'll go higher because I know the lead is heavier.' Ey also predicted that the lead would cause a greater rise in level than the red cylinder, and having discovered that it did not, he nevertheless went on to assert that the lead would lift the water higher than the black cylinder. He even went so far as to justify the second prediction by saying: 'It'll be as I said before', and went on to ignore his own observations when he repeated his mistake once again with the orange cylinder! Now, either he was quite unable to argue logically, or else he must have believed that experience is an unreliable guide. In fact, both possibilities reduce to what we have called a failure in composition. Thus, Bra who had granted that $R = Pb$ and who remembered perfectly well that $R = B$ 'because they're the same size', nevertheless refused to conclude that $Pb = B$: 'The lead will make the water rise higher because it's heavier.'

Compositions involving the clay coil and the three cylinders are equally mistaken, not surprisingly so, since in their case the child must guard not only against the illusion that the objects he is shown differ in weight (the lead cylinder has the same shape and dimensions as the aluminium cylinders) but also against the belief that they differ in weight and volume combined: the clay coil looks larger but lighter than the cylinder. Thus Ey expected that the 'sausage' would make the water rise higher than the red cylinder because 'it is bigger' and, having discovered the equivalence, he nevertheless failed to conclude that it would produce the same effect as the black cylinder; he predicted that the water level would rise more with the sausage 'because it is thicker'. The demonstrator then drew his attention to the composition $CL = R$, $R = B$, whence $CL = B$; but Ey refused to grant that conclusion: 'The water will rise more with the sausage.' Hae reacted to the clay much as he had to the lead; having granted that $CL = R$ and being asked whether $CL = B$, he said: 'Also the same. No, this one (B) will make the water rise higher because it's heavier.' After the experiment ($CL = B$), though recalling that $Pb = B$, he refused to grant that $CL = Pb$:

'The lead will make the water rise higher because it's heavier.' Clearly, these equalizations, let us not yet say of the volumes, but of the volumes × the weights reflect precisely the same failure in composition as Stage I responses to weight determinations. None of our subjects proved capable of deductive operations even by the transitive equalization of three like terms, and this because the necessary compositions come up against one of their pre-relations: they accept the evidence of their eyes in the singular case they have just examined, but fail to extend it to identical cases.

A fortiori, they cannot effect additive compositions with homogeneous or heterogeneous objects. Here, the experiment with the large cylinder (III) is particularly instructive because, unlike the bars we used for the composition of weights and which necessarily reduced to the selection of entire units, the cylinders, etc. force the child to produce his own graduations and hence to disclose what he thinks about the relationship of the parts to the whole. Now, our subjects are so incapable of composition that, having indicated the difference in level caused by the immersion of one of the three small cylinders, they cannot predict that two cylinders will produce twice that difference, or that the large cylinder will displace as much water as three small cylinders. Thus Ey ignored all the proportions and Bra pointed to much too low a level when the second small cylinder was added to the first, and, having discovered his mistake, to much too high a level when the third cylinder was added. Moreover, he claimed that the three small cylinders would displace more water than the large one 'because there are three of them'. Having seen his mistake once again, he nevertheless went on to claim that the 'long one' will 'perhaps' make the water rise higher 'because it's long'.

Mag reasoned in much the same way when he said that 'the water will rise higher over here because this one (III) is longer', and when, having discovered the equivalence, he nevertheless refused to grant that it would be conserved if the small cylinders were stacked horizontally.

The addition of heterogeneous objects produces precisely the same results. If an aluminium cylinder is combined with a lead cylinder or with a coil of clay and compared with two cylinders in the other vessel, the jar containing the lead or the clay is said to cause a greater rise in level 'because the lead is heavier' or 'because the clay is bigger'.

If these are the only kinds of compositions children at Stage I can produce, it goes without saying that they cannot possibly elaborate the law relating the rise in water level to the volume of the submerged object. This is because the two are not only interrelated but also constitute the two sides of one coin. Logical composition, in fact, depends on at least two factors : (1) the adoption of notions of relations that lend themselves to composition, i.e. that are non-contradictory and lead to non-contradictory conclusions; and (2) the ability to arrive at conclusions by deductive paths, i.e. by reversible operations. Similarly the elaboration of a law depends on at least two factors : (1) the adoption of objective notions or relations, i.e. of notions that reflect reality rather than subjective or egocentric impressions; and (2) the realization that experimental results, unlike phenomenalistic influences, are governed by strict laws. It follows that the two obstacles in the path of the compositions under review are precisely the same that stand in the way of the discovery of the displacement law : the child starts out with a global notion (weight × quantity of matter × volume) which is, in fact, a prerelation, and as such neither composable from the logical point of view nor objective from the factual point of view. Moreover, he has no confidence in deduction and keeps reverting with each new problem to his preconceptions and phenomenalistic approach, which prevent his learning anything from the experiments he himself has just made.

Let us deal with the prerelations first. As we saw in Chapters 1–9, children at Stage I explain the rise in level by arguing that the weight of an object is proportional to its apparent quantity of matter and volume, and also that weight is an active and substantial force, which, as Hae put it, 'goes to the bottom and makes the water rise'. This undifferentiated force is generally thought to be reflected in such qualities as 'thickness' or 'length' and, what is more, in a global form that does not lend itself to logical composition. Thus 'thickness' is used to refer to the various respects in which a particular object is superior to others, and precisely because these aspects vary the child is drawn into systematic equivocation. In particular, whenever an object is superior in one respect, the child readily concludes that it must be superior in others. Thus, Ey argued that a vertical cylinder is 'thicker' than an identical cylinder lying on its side; that a lead cylinder is 'thicker' than an aluminium cylinder of the same dimensions, and that a coil of clay is 'thicker' than a metal cylinder. By 'thicker' he therefore referred variously to the volume, the weight or the length, which is awkward enough; worse still, this usage makes it impossible to tell whether he believed that the larger its volume the heavier a body must be, or *vice versa*, e.g. whether it is the volume or the weight that pushes the liquid up. In any case, it is clear that this conception of thickness is an egocentric-*cum*-phenomenalistic prerelation, for while experiment suggests that the weight is proportional to the volume it also shows that it need not be. In other words the concept used by our subjects constitutes a global schema·of heaviness, strength and thickness; it is therefore both muscular and visual (and partly social inasmuch as the weight and the size of people plays a large part in the child's daily life). Now, if this prerelation does not lend itself to logical composition, it goes without saying that it cannot form the basis of an objective relationship allowing of the elaboration of the displacement law, and this for the simple reason that the child is forced to speak in two contradictory languages: by 'thickness' he refers alternately to weight and volume (*cf.* Ey). But even those who do not use this term explicitly are forced, when speaking in terms of weight or volume, to employ one and the same preconception to both systems. Thus Hae predicted that one of the cylinders would raise the level more than another 'because it is heavier', and seeing that it did not, he nevertheless went on to assert that if the second cylinder were placed in a vertical position it would produce a stronger effect, and this because 'it takes up more room ... it is bigger'. Bra, for his part, first invoked the size as the decisive factor, then the length and the weight. Now, quite obviously, if these equivocations mean that a particular relation can be accepted or rejected at will, they also mean that the water can rise to the same level for two entirely different reasons: the immersion of a large but light object, or the immersion of a small but heavy object.

The second factor of the elaboration of the displacement law, i.e. the victory of experimental regularity over the fleeting appearances, eludes children at this level as well. For not only do they prove quite incapable of drawing logical conclusions from the data of a possible composition, but they also hold experiment in utter contempt: the result of a particular experiment tells them nothing at all about the next. When Ey, for example, predicted that the lead cylinder would make the water rise higher than the red cylinder, and when the experiment proved him wrong, he nevertheless went on to say of the black and lead cylinders: 'it'll be as I said before', meaning that the lead would this time prove him right, i.e. that it would make the water rise higher precisely because this had not happened the first time. Now, this response not only reveals a failure to draw the logical conclusion $Pb = B$ from the premises $Pb = R$ and $R = B$, but also a strange impermeability to experience. Ey fully expected that the second test would bear out the very hypothesis the first test had invalidated. Moreover, proved wrong once again he fully expected that a third test would vindicate him at last!

We encountered the same phenomenon in Chapter 4: children

at Stage I who have not yet grasped the conservation of substance may discover that the level of the water does not drop after the dissolution of the sugar and that the weight remains unchanged, but they nevertheless fail to conclude that their original hypotheses were wrong.

All in all, therefore, at Stage I lack of composition and deduction go hand in hand with the primacy of phenomenalism and with the inability to learn from experience. This is why failure in logico-arithmetical and physical composition is necessarily associated with failure to elaborate the displacement law: in fact, the deductive equivalent of the inductive or experimental components of non-composition is nothing other than deafness to the lessons of experience.

§2. Stage II : the transductive beginnings of composition and of experimental regularity

Stage II sees the beginning of the composition of simple equalities by transductive analogy, and also the dissociation of weight from volume. As a result the child can begin to elaborate the displacement law. Here are some examples :

RAM (6; 11) says the cylinder O will make the water rise *'as high'* as N *because it's as heavy.*—And these (R and O)?—*The same.*—And these (R and B)?—*The same.*—And if I put this one (R) on top of that one (B)?—*It'll rise up to here* (points to too high a level, then corrects himself).—And (R + O + B)?—(Roughly correct.)—And what about the big one (III) in the other flask?—*The water will rise higher because it's heavier than the three together. Oh no, it'll rise the same because it's as heavy.*

'What about this one (BR = a brass cylinder) and that one (B)?—*The water will rise higher because this one* (BR) *is heavier than the rest.*—Look.' (Experiment.)—'*Oh, it's the same.*—And what if I put (R + O) in one glass and (BR + B) into the other?—*It'll rise the same. Oh no, not the same because* (BR + B) *has more weight.*—And how far will it rise with this one (BR)?—*A little higher.*—And BR against B?—*The same.*—And (BR + B) against (R + O)?—*The same.*'—(Experiment.) 'Why?—*Even if one of them is a bit heavier it's the same.*

'What happens if I put this one (Pb) in one flask and these (B + R) in the other?—*The lead is heavier than the two together. These two* (B + R) *are lighter. The water won't rise so high.*—And the lead with the black one alone?—*The water will rise much higher with the lead.* (Experiment.) *Oh, it's the same.*—And BR with B?—*We just saw it's the same.*—And Pb with BR?—*The lead will make the water rise higher, because it is heavier.*—But Pb with B?—*The same.*—And B with BR?—*Also.*—And Pb with BR?—*Ah, the same.*—Why?—*They're both heavy.*—Which is heavier?—*The lead, but it makes the water rise the same.*—Why?—*The other one is heavy as well.*

'And this (CL) with B?—*This one* (B) *will make it go higher.*—Why?— *Because the clay is lighter, so the water will rise less.*—Look.' (Experiment.)— '*The same thing.*—And the lead with B?—*We just saw, it's the same.*—And BR with Pb?—*The same.*—And CL with Pb?—*It'll go higher with the lead.*— Look.' (Experiment.)—'*It's the same.*—And (B + Pb) with BR?—*The same.*— Why?—...—But how did you get that?—...—And Pb with CL?—*The water will rise more with the lead, because the clay is lighter.'*

PEL (6; 10). 'R and B?—*They'll make the water rise the same because they are the same size.*—And if R is laid on its side and B is stood upright?— *The black one will make the water rise more because it pushes on both sides* (points to the height) *and that makes it heavier* (!).—And B and R standing on end?—*It'll rise the same, they're the same length.*

'And Pb with O?—*The lead will make the water rise more because it's heavier.* (Experiment.) *Oh, it's the same!*—And O and B?—*The same as well.*—And Pb with B?—*The lead will make the water rise more, it's heavier.*

'And (R + O + B) with III?—*The big one will make the water rise more, because it's heavier than the three.*—And how about these (the three Is superposed) and that one (III horizontal)?—*It'll be the same because they're the same size.*—And (the three Is upright) and (III horizontal)?—*The big one will make the water rise more, because it is heavier.'*

GRA (7; 0). 'O and R?—*The same.*—B and R?—*It'll be the same because they're the same size.*—And O with B?—*Not the same, because they aren't the same size.'* Looks them over again. '*Oh, yes. They will be the same as well.*—And (B horizontal) with (O vertical)?—*The water will rise more over here* (O).—See for yourself.' (Experiment.)—'*Oh, it's the same.*—Why?— *They're the same size.*

'And O with Pb?—*The lead is heavier.*—And this one (O)?—*It's a little lighter.*—So?—*The water will rise as much, because they're the same size.*— Pb and B?—*Also, because they're the same size.*—And (O + N) with (R + Pb)?—*The water will rise more over here* (R + Pb).—Why?—*It's heavier.*— Look.' (Experiment.)—'*Oh, it's the same because they're the same size and the same weight*(!).—And these (Pb and R)?—*The lead will make the water rise more, it's heavier.*—Look.' (Experiment.)—'*It's the same.*—And (O + B) with (Pb + R)?—*The heavier lot* (Pb + R) *will make the water rise higher.'*

R will make the water rise higher than **CL** '*because it's heavier'.*— (Experiment.) 'And what about B with CL?—*It's not the same.*—Why not?— *Because* (BR) *is heavier.* (Experiment.) *Oh, I see!*—And Pb with CL?—*It'll go higher with the lead.'*

CAN (7; 9). 'R and O?—*It'll be the same.'* He weighs them in his hand. 'And these (R + B)?—*They'll make the water rise the same.*—Which ones have we tried?—*These* (R and O) *and those* (R and B).—And what about these (B and O)?' He refuses to commit himself and makes the experiment. 'And what about these B upright with O horizontal)?—*I don't think it's the same because when it's up it makes the water rise a little more; it's got more force when it's heavy.*

'And this one (Pb) with that (R)?—*More with the lead because it's heavier.*— Look.' (Experiment.)—'*It's the same, isn't that odd!*—And Pb with O?— *It's the same as with those* (Pb and R).—And (O horizontal) with (Pb vertical)?—*It's the same; they've got the same weight all the time* (!).

'And what if I put (Pb + O) with (R + B)?—*It's the same. No, it won't weigh the same; the water will rise higher with* (Pb + O) *because one is made of lead; it's thicker and it's got more force.*'

CHRI (8; 0). 'What will happen if I drop this one (R) into the water?—*It's going to rise. When there's something in the water it has to rise because the water has got to go somewhere* (= displaced volume).—And what if I put this one (B) in?—*Precisely the same; they are the same size.*—Very well. Now you remember that. What about O with R?—*Also the same.*—And O with B?—*I'll have to see if they're the same size.*' Measurement. '*Yes, they'll make the water rise the same.*—What about this one (O horizontal)?—*I don't know. Just wait, it seems to look smaller, but it must still be the same. Perhaps the water will rise as much.*

'And Pb with R?—*The lead is very heavy, so the water will rise more, I think.*—Try it'.—(Experiment.) '*It's the same.*—And Pb with BR?—*The lead is heavier, so the water will rise higher with it. Oh, no; I'll have to look at the size. It's the same size, I think.*—And Pb with CL?—*The clay isn't so heavy.*' He drops it in the water. '*It's the same.*—And the clay with the brass?—*The clay looks smaller and it's lighter.*—What do you think?—*Perhaps it'll rise more over here* (BR)?—Why do you think so?—*I don't know. In any case it's not the same size. I just don't know.*'

To dispel any possible misunderstanding, we would point out, first of all, that though some of these children start out with expressions taken from the language of size they have not necessarily differentiated the volume from the weight only to revert to the confused ideas associated with Stage I in the course of the experiment. Rather must we take it that their initial notion is still relatively undifferentiated: differentiation develops with simple compositions and their experimental verification but regresses with compositions involving different weights, in which case the opposition between volume and weight becomes too confusing.

The result is a continuous series of transitions. Thus while some subjects such as Ram and Can started with the idea that the rise in level is proportional to the weight and not to the volume of the immersed object, others such as Pel invoked 'the same size', but in the belief that the weight is more or less proportional to the volume, for we need merely stand one of the two equal cylinders on end for the primacy of the weight to assert itself: 'The black cylinder pushes on both sides and that makes it heavier' (Pel). Gra used much the same argument, but when he discovered that the vertical and the horizontal cylinders produced identical effects he began to differentiate the weight from the volume. However, with the composition of (O + N) with (R + Pb), he resorted to the weight of the lead, and when the experiment once again undeceived him he reverted to so complete a lack of differentiation that, to justify the equality he had just discovered, he concluded against all the evidence of his senses that the two pairs were 'the same size and the same weight'! Chri clearly went further than all the rest in his differentiation of the weight from the volume because he invoked the volume to offset the illusive appearance of the horizontal cylinder and because, when asked to compare the aluminium with the brass cylinder, he said, 'I'll have to look at the size.' But he, too, reverted to the weight when presented with the lead cylinder, and found himself at a complete loss with the coil of clay.

We can now look at the actual composition of these children. In general, all of them try to base their predictions on what they have just discovered experimentally, and to that end they look for a common term: thickness, size, weight or volume. But their deductions still lack logical rigour and hence cannot be called true compositions; they are still based on analogy or on transduction (no operational reversibility).

In fact, even with objects of homogeneous shape and weight, though they can successfully compose the weights $(A = A') + (A' = A'') = (A = A'')$ because they rely on the quantity of matter, they do not, in fact, go beyond Stage I: they do not deduce $A = A''$ but arrive at it by analogy. Thus, Ram said that $B = O$, $R = B$, hence $O = R$, not because $O = R$ follows necessarily from the first two equalities but because all the terms were similar. The same approach was also used by Gra who said that $O = R$ and $B = R$ 'because they are the same size', but went on to assert that O and B 'aren't the same size' and then changed his mind but purely for the sake of uniformity, looked them over again, and said: 'Oh, yes, they'll be the same as well.' Chri and Can also failed to treat $A = A''$ as a necessary consequence of $A = A'$ and $A' = A''$. 'I'll have to see if they're the same size,' said Chri of $O = B$, after having stated that $R = B$ and $O = R$.

In respect of equivalences with several homogeneous objects and one heterogeneous object two points must be specially noted. First of all, our subjects, like those at Stage I but to a lesser extent, tend to forget what equivalences they have established as soon as the lead and the brass cylinders are introduced, when they revert to the idea that the change in level is due solely to the weight of the immersed object. But, unlike subjects at Stage I, as soon as they discover that the first two heterogeneous objects produce the same effect, they tend to extend the principle and hence begin to dissociate the weight from the volume. Thus, Ram thought first that the lead would make the water rise higher than $O + R$, the brass higher than B, and the clay less than B. Pel and Can thought that the heavier lead would have a greater effect than the aluminium; and even Chri, who resisted the impression of the brass, gave way to the weight of the lead, if with some reservation. Gra was the only one to equate the effects of the lead with that of the other cylinders straightaway. However, all our subjects except for Ram and Pel, who produced intermediate responses, learned from the experiments. Thus Can, having discovered that $Pb = R$, said of Pb and O: 'It's the same as with those two (Pb and R)', and even realized that the equivalence is conserved if one of the two cylinders is laid on its side. Chri, for his part, having discovered that $Pb = R$, said of the lead and the brass: 'The water will rise higher with it (Pb). Oh, no, I'll have to look at the size.'

The second point worth mentioning is that when both the weight and the shape of the objects differ (and not merely the weight as with the brass and the lead cylinders) the equivalences do not even give rise to analogic compositions or to inductive predictions. With the coil of clay, for example, no equivalences were thought to be transitive and no lessons were drawn from the experiments. Thus, when Ram discovered that $CL = B$, after predicting that the clay, being lighter would make the water rise less, and remembering that $B = Pb$ and even that $BR = Pb$, he still refused to conclude that the clay is equivalent to the lead. When he found that it was, he arrived at $(B + Pb) = (CL + BR)$ but by sheer guesswork, and after discovering that he was correct, he nevertheless denied that $CL = Pb$! Gra, who established the correct equivalences with the lead, also failed with the clay. When he found that, contrary to his prediction, $CL = B$, he, too, refused to conclude that $CL = Pb$. Even Chri, the most advanced of these subjects, who claimed that $Pb = BR$ because of the 'size' and who discovered that $CL = Pb$, was at a complete loss when it came to the clay and the brass: 'I just don't know. In any case it's not the same size. I just don't know!'

The contrast between these two types of reaction is extremely instructive. Clearly, the only equivalences to concern the volume as such are those bearing on objects of different shape and weight. Now such (operational) equivalences are not yet composed in any way by children at Stage II, who begin to dissociate the weight from the volume only in the case of objects of identical shape but of different weight. As a result they begin to compose what equivalences they have discovered by reference to the 'size' (cf. Chri's response), i.e. to the apparent quantity of matter. We found the same reaction in Chapter II: subjects at Stage II proved successful with compositions of bars of the same dimensions but not with bars of the same weight and of different shapes. We must also bear in mind that, even in respect of the size, subjects at this stage produce analogic and transductive, not logically necessary constructions:

having discovered to their surprise that the rise in level depends on the size instead of the weight of the immersed object, they bow to experience but cannot explain what has been happening. Hence when the volume alone intervenes, as with the coil of clay, they refuse to produce any compositions, even of an analogic type.

When it comes to additive compositions, some subjects at this stage are successful with the homogeneous cylinders $R + O + B = III$ but not consistently so, and only after hesitations. Thus Ram pointed to roughly the correct level for two or three cylinders of the order I, then claimed that the cylinder III would produce a greater rise in level because it is 'heavier than the three together', and finally came down in favour of the equivalence because 'it is as heavy'. Pel, for his part, did not grant the equivalence except in one single case, on the grounds that 'the big one is heavier than the three'.

As for additive compositions involving heterogeneous elements, they are systematically resisted, which leads to a resurrection of the prerelation weight × volume. Thus, Ram thought that $(BR + B)$ would make the water rise higher than $(R + O)$ because it 'has more weight', and this precisely after he had discovered that $BR = B$. Similarly, Gra claimed that $(R + Pb)$ would produce a greater effect than $(O + B)$, although he had established all the equivalences. Even when he was shown that the lead did not make the water rise more than the coloured cylinders, he went on to claim that it did when the objects were presented to him in pairs. Can, who had just said of the lead and the other cylinders 'they've got the same weight all the time' (what he meant was the same lifting force), nevertheless claimed that $(Pb + O)$ has a greater effect than $(B + R)$ because 'it's made of lead, it's thicker and has more force'.

Such are the main reactions at this stage. To appreciate the unity behind their baffling diversity, we must try to see the problem through the child's own eyes. When we asked him to predict the weight of an object after he had correlated two others (Chapter 11) the problem was very simple; he was familiar with the concept he was asked to employ. In the present series of tests, on the contrary, he is told to compose volumes without realizing that this is what he is expected to do! He is asked to answer a concrete question: if A causes the same rise in level as A', and A' causes the same rise as A", will A cause the same rise as A" or not? To solve it he need merely appeal to logic; from the equivalences he has established he can easily tell that only the volumes are concerned so that, even if he knows nothing of the law of levels, he can readily deduce it from the logical compositions. Now, what is so striking, and moreover resounds to the child's great credit, is that he refuses to deduce $A = A''$ from $A = A'$ and $A' = A''$ while he has not yet fully grasped why the level rises. And since he believes that the rise in level depends on the 'thickness' of the immersed bodies (a term he applies indifferently to their weight and volume), the use he makes of the formal equivalences and his eventual discovery of the experimental law will depend on the manner in which he replaces this non-composable prerelation with composable and differentiated concepts of volume and weight.

In short, children at Stage II make a start with analogies, though not yet with deductive composition: they are uncertain of their conclusions because they still fail to effect a complete dissociation of the volume from the weight. They also begin to grasp the idea of experimental regularity, but by induction only; and their inductions are often unsure and hampered by the relative failure to dissociate the weight from the volume in their original 'thickness' schema.

This explains why they are able to establish simple equivalences when presented with homogeneous cylinders: $(B = R) + (R = O) = (B = O)$, but transductively, not deductively: they are too uncertain of the concrete effects of the immersed objects on the water to rely on deductive arguments, however elementary. For that very reason they deny the equivalences as soon as the position of one of the cylinders is changed. They are somewhat more successful in establishing equivalences with objects of the same shape and differing in weight only (lead and brass cylinders), which may seem paradoxical but is easily explained by the fact that, having discovered that the weight must be dissociated from the 'size', they focus their attention on the latter alone, and since 'size' is a perceptible and quite obvious quality of matter, they can argue more correctly about it. By contrast, when their compositions bear on objects differing in both shape and weight, they fail completely since, in that case, they must first isolate the volume as such. When it comes to additive compositions these children, needless to say, resurrect all the prerelations they have partly vanquished, and this by means of a mechanism we have met before (see Chapter 11). The only additive compositions some of them can produce are between the three cylinders I and the large cylinder III; but here again they proceed by analogy and are thrown off course by changes in the position of any one of the cylinders.

A parallel situation occurs in the discovery of the displacement law. As the child begins to effect certain compositions, be it only by analogy or transduction and not yet by deduction, he also learns to defer to experience. When Can and Chri, for instance, observed that the lead cylinder, though much heavier than the aluminium, did not have a greater effect on the water, they concluded that all the other aluminium cylinders would act in the same way, thus evincing an incipient grasp of logic and of experimental regularity. Now it is the fusion of these two which leads to the construction of the displacement law. However, the regularity is not yet based on a principle of induction extending to all the relations established by the experiments: it applies exclusively to objects of similar shape, and even then without certainty, and not to objects of heterogeneous shape or to additive compositions.

§3. Stage III: the emergence of deduction and the inductive discovery of the displacement law

Stage III marks a decisive step forward: the emergence of logico-arithmetical deduction reflected in the use of such terms as 'because', 'therefore', 'of course', etc. At first these logical compositions are applied exclusively to the weight, which bears out what we have said about Stage III responses in the last chapter. But, precisely because they are rigorous, and no longer analogic, these compositions pave the way for a much more precise dissociation of the weight from the volume and hence for a definite break with the 'thickness' schema. In particular, the child learns to appeal to the volume as such and hence gradually discovers the displacement law. However, this discovery is only made in the wake of the experimental findings; it remains purely inductive and does not yet give rise to the direct deductions associated with Stage IV.

Here are some examples:

DET (7; 2). 'What will happen if I drop this cylinder (B) into the water?— *The water will rise up to here, because the cylinder takes up room.*—And R?— *The same.*—And B and O?' He looks carefully. *'The same as well.*—And R and O?—*Yes, they're the same size.*—And Pb with O?—*The lead is heavier, so the water will rise up to here* (higher level).—And with R?' He points to a slightly lower level. Experiment with Pb and O. *'Oh, it's the same.*—And Pb with B?—*It's the same, because* (Pb) *weighs the same as* (R) *so they're the same as this one* (Pb) *with that one* (O).

'And what about (O + Pb) with (R + B)?—*The first two* (O + Pb) *will make the water rise higher. Oh no, they'll make it rise the same, it doesn't matter that they're heavier.*—And (BR + Pb) with (O + B)?—*It makes no difference that they're heavy; it's the same as if they were like these* (O + B).

'And what about this clay?—*It'll take up less room, there's less of it.* (Experiment.) *Oh, it's the same as the rest.*—As which rest?—*All of these* (cylinders).—And (CL + Pb) with (R + B)?—*It's the same, because they weigh the same. Oh, no; it's because they take up as much room in the water as these two* (R + B).

'If I drop this one (cylinder III) in, what must I put in the other flask to get the same level?—*If you put this one* (III) *in, the water will rise less because it's big and thin: to get the same level you'd have to put in these two* (B + Pb).— Why?—' He picks up cylinder III, compares its length and height with those of cylinder I, then weighs III against Pb in his hands and says: '*Yes, they weigh the same.*—But will they take up the same space?—*Yes,* (B + Pb) *will take up the same room as that one* (III).—The same room?—*Yes, they weigh the same so they take up the same room (!).*—Try it.'—(Experiment.) '*Oh no, we need more* (puts in BR + CL + B).—Why?—*It doesn't matter which, these three* (BR + Pb + B) *take up as much room as that one* (III).— Why?—*They weigh the same.*' He weighs them. '*Oh, no, they're going to make the water rise more. No, I'm wrong, they'll take up the same room.*' He thus identifies III with any three cylinders of order I.

BON (8; 5). '*If you drop something in the water, it causes pressure and so the water will rise.*—R and B?—*Same thing, they are the same weight.*— And O with B?' He weighs and measures the lengths. '*The same.*—And O and R?—*Yes, they're the same length, the same width and the same weight.*

'And Pb and O?—*The same.*—Why?—*They're not as heavy, but they're as tall as each other.*—But what does the weight do?—*It makes a slight difference.*—But you have just said they're all the same.—*Yes.* (Experiment.) *They're the same, it's as if we put* (O) *in with* (B).—Why is it like that?— *We saw that* (B) *is the same as* (O) *so I say it's as if we put it Pb in with* (B).

'And (Pb + O) with R + B)?—*The same. Perhaps not. The lead is heavier. The water will rise higher; the lead makes more pressure when it gets into the water.*—You saw Pb and O, Pb and B, and B and R?—*Yes, they were the same.*—And if I put in (Pb + R) and (O + B)?—*The water will rise as high. We've tried it with all of them, and the three made the water rise as high as the lead.*—Are they the same weight?—*Perhaps the lead will make the water rise higher because it's heavier.*'

He thinks that CL will make the water rise less than O. (Experiment.)— 'Why do they do the same?—*They have the same pressure even though they aren't the same weight. It's because they are the same height and the same width. If we turned the clay into a ball and the ball into one piece like that* (the cylinder) *it would be the same thickness and the same height.*'

'And what about (CL + Pb + BR) and (R + O + B)?—*That will make the same height. If we turned all these* (the second set) *into a ball, they would be lighter but they would have the same pressure, the same height and the same size, so the water would rise as high.*'

RENT (8; 10). 'What would happen if I dropped this one (B) in?—*It would make the water rise.*—Why?—*Because it's got weight and takes up room.*— And R with O?—*The same.*—And B and O?—*Yes.*—A and B and R?— *For sure.*—And Pb?—*That'll make the water rise a little higher; it's heavy and takes up more room. Oh, no; the same room, but it's heavier so it makes the water rise more.*'—(Experiment with R.) 'And what about B and O?— *They'll do the same.*'

'And (Pb + B) with (R + O)?—*The water will rise higher on the first side, because it's heavier.*—But what about Pb and R?—*Oh, yes; they're the same, so it'll rise the same as with* (R + O), *because they take up the same room even though they're not the same weight.*—And what if I lay this one (Pb) on its side?—*It's the same, because it's still the same weight and still takes up the same room.*

'And CL with B?—*Perhaps the same.*' (Experiment.)—'And with R?—*The same, because it still takes up the same room.*—And what if I changed the clay into a cylinder?—*It would be bigger because it's lighter. Oh no, it takes up the same room, it's the same.*—And if I turned the clay into a disc the width of the glass?—*The water will rise just as much, because the disc will take up the same room.*

'What if we pour the water that has risen up into a glass cylinder?—*We'd need a larger cylinder than this one* (B), *because there is more water than the weight of this metal cylinder.*—Why?—*If it's the same size, only part of the water will go in.*—How much room does the cylinder take up in the water?—*That much* (points to the difference in level).—So how big must the glass be?—*Bigger, because there's more water.*—What counts, the weight or the room it takes up?—*The room! Oh, it's the same. The glass will have to be like the cylinder!*'

CLAN (7; 11). 'What about B and R?—*They'll make the water rise the same, because they are the same weight.*—And B and O?—*Perhaps it'll be the same as well.* (Experiment.) *Yes, it is.*—And what about R and O?—*The same as well. They weigh the same as the black, so they must be as heavy as each other, and so the water will rise as high.*—And these two (R and Pb)?—*The lead will lift the water higher.* (Experiment.) *Oh, they don't feel the same weight, but in the water they all go to the bottom, so it'll rise as high.*—And Pb with B?—*The red weighs as much as the black, so the lead and the black will do the same to the water.*—And (O + B) with (Pb + R)?—*These two* (O + B) *will weigh the same, and those two* (Pb + R) *don't weigh the same; but it makes no difference, the lead and the orange will both drop to the bottom and the water will rise the same.*—And (BR + Pb) with (O + B)?' He hesitates. '*The brass and the lead will make the water rise higher, because they are both heavy.*' But in the end he equates their effects.

KEL (7; 11). 'What if I drop this one (R) into the water?—*It'll drop to the bottom and the water will rise.*—Why?—*The water stays where it is when nothing is dropped in, so if we drop something in it must rise.*—And this one (O)?' He weighs it and compares its height. '*Same thing.*—And O and B?—*Let's see the weight.* (Weighing.) *It's the same.*—And R and B?—*We've tried them with the others. The water will rise as high, because the weight is the same.*—And Pb with R?—*The lead will take up more room, it won't be the same.* (Experiment.) *Oh yes, it is.*—And Pb and B?—*Also; because we saw that with the lead and the red. The red and the black are the same, so the lead and the black must be the same as well.*

'And (R + O) with (Pb + B)?' He hesitates. '*The red and the orange will make it rise less. Oh no, the water will rise as high as with the lead and the black.*—And BR and Pb?'—(Weighing.) '*The lead will make it rise higher.*'—(Experiment.)—'And (Pb + BR) and (R + O)?—*We saw that* (Pb) *and* (O) *make the water rise as high. Now* (Pb) *is like* (BR) *so* (BR) *can go equally well with that one.*' The demonstrator changes BR and R around. '*Still the* same. *We know that this one* (BR) *is the same as that one* (R). *We tried it with* (Pb) *and* (BR), *and with* (Pb) *and* (R).

'And what about the clay and this one (O)?—*It's the same, because they're the same weight.* (Weighing.) *Oh no.*' He drops them into the water. '*They are not the same weight, but the water still rises as high.*—And what about CL and Pb?—*The lead weighs more than* (R) *but even though it's heavier it makes the water rise the same, so the clay will also make it rise the same.*—And what about (O + BR) with (CL + Pb)?—*That one will weigh more, but it's going to make the water rise the same because this one* (CL) *is the same as that one* (O), *so even if* (O) *was heavier the water would rise as high.*

'And what must be put in to get the same rise as with that one (III)? Will it depend on the weight or on the room it takes up?—*It's the room that matters.*' He superposes Pb + BR + R. 'And what if we put one cylinder upright and the others lying on their sides?—*It makes no difference; the water will rise as high.*'

These reactions are of the utmost importance because they show by what paths these subjects arrive at the discovery of the displacement law.

Let us note, first of all, that none of them came to the test with a grasp of the law, and that, at the outset, all of them still adhered more or less firmly to the 'thickness' schema, i.e. they believed that the 'thicker' a solid the greater the pressure it exerts on the water, and that once it has reached the bottom its weight, not its volume, continues to keep the level up. Thus Kel explained that, when the cylinder drops to the bottom, 'the water will rise'. Moreover, he claimed that at equal volumes, the heavier of two objects causes a greater rise in level because it displaces more water: 'The lead will take up more room!' And Bon said that the lead plus the aluminium cylinders would produce a greater effect than two aluminium cylinders because the lead, being heavier, exerts 'more pressure when it gets into the water'. Even Det, who said right at the beginning that 'the water will rise up to here, because the cylinder takes up room', went on to assert that, at equal volumes, the lead would have a greater effect than the aluminium, and even went so far as to identify the cylinder III with only two small ones on the grounds that 'they weigh the same so they take up the same room'!

How then do these children discard their prerelations and arrive at the correct law? By and large, they proceed as follows: they start out with rigorous compositions of equivalent weights (which, as we saw in Chapter 11, they can readily do). Next, when the experiments prove them wrong and they realize that the equivalences are independent of the weight, they conclude that the weight 'makes no difference', and hence gradually learn to dissociate the weight from the volume, and to base their arguments on the latter.

Their composition of simple equivalences shows that their approach is quite different from that used by children at earlier stages: they no longer rely on analogies but apply strict deduction, at least in elementary cases. Thus Clan said straightaway that the red and the orange cylinders produced equal effects because 'they weigh the same as the black, so they must be as heavy as each other, and so the water will rise as high'. Kel, having first compared the red and the black to the orange, also concluded that they must be the same because 'we've tried them with the others. The water will rise as high, because the weight is the same,' etc. Now, these compositions introduce no new problem because they are all based on weight; the only novelty is the deductive rigour, but since the latter is characteristic of Stage III weight compositions in general, it is only to be expected that it should also be applied to the weight of submerged solids.

It is only when they are asked to compose equivalences with objects of the same shape but of different weight that they are suddenly forced to distinguish the weight from the volume. Thus, Clan expected that the lead would have a greater effect than the red, but seeing that it did not, he exclaimed, 'Oh, they don't feel the same weight, but in the water they all go to the bottom, so it'll rise as high.' Next he concluded that the lead must also be equivalent to the black and, though he still invoked the weight, he disregarded it in practice: 'The red weighs as much as the black, so the lead and the black will do the same to the water.' Bon even went so far as to predict that the weight of the lead would play no part at all: the orange cylinder and the lead, he said, 'are not as heavy, but they're as tall as each other'. Admittedly he added at once that the weight does make 'a slight difference', which shows how difficult he must have found the dissociation; but he nevertheless concluded that the lead cylinder behaved 'as if' it were an aluminium cylinder of the same dimensions.

We see what great strides these children have made: at Stage II equivalences as between objects of the same shape but of different weights lead merely to a dissociation, within the general global schema, of the relative weight from the size or quantity of matter, and the child still attributes a greater effect to the heavier object; at Stage III, by contrast, he has advanced to more rigorous weight compositions (see Chapters 9, 10 and 11); he realizes at once that the weight plays no part in the displacements, and hence tries to disregard it. Thus Det said of (O + Pb) and (R + B): 'They'll make it rise the same, it doesn't matter that they (O + Pb) are heavier ... it's the same as if they were like these (O + B).' Similarly Clan said that 'those two don't weigh the same; but it makes no difference'.

And Kel: 'The lead weighs more (than R) but even though it's heavier it makes the water rise the same.'

But why is it that, though these children disregard the weight, they can nevertheless go on to produce compositions of such great rigour (in any case, as far as simple equivalences are concerned)? The answer is quite simple: having discovered the equality of the levels, they go on to compose $(A = A') = (A' = A'') = (A = A'')$ in the belief that they are still dealing with 'thicknesses' based on weight. Then, finding that the weight plays no part, and that it can be omitted from their compositions,[1] they preserve the equivalences and look for a new factor by which all the elements can be equalized.

Now their responses to the coil of clay provide us with the clearest illustration of the progress they have made (at Stage II the same test produced wholly negative results). Thus Det thought, first of all, that the clay would raise the level less because 'it will take up less room, there's less of it', thus relying on size or quantity of matter rather than on weight, and having discovered the equivalence of level he concluded: 'It's the same because they (CL + Pb) weigh the same. Oh no; it's because they take up as much room in the water as these two (R and B).' In other words, he was considering the 'room' the immersed objects occupy regardless of their shape and *a fortiori* of their weight, i.e. he had implicitly discovered the role of their volume. Bon, for his part, was quite explicit: discovering that the coil was equivalent to the cylinder O he at once constructed a geometrical representation: 'They have the same pressure even though they aren't the same weight. It's because they are the same height and the same width. If we turned the clay into a ball and the ball into one piece like that (the cylinder), it would be the same thickness (diameter) and the same height.' Rent, by contrast, who also realized that the coil could occupy the same space as the metal cylinder despite their differences in shape, failed to equate them at first. Instead he reverted to the weight, and maintained that the clay cylinder would have to be 'bigger' because it was lighter. A moment later, however, he corrected himself and said that a clay disc of the same diameter as the glass vessel would have to be of the same height as the column of water it displaced.

But, though these children can equate the volume of the clay with that of a metal cylinder, they do not as yet grasp that the volume of the displaced water can be treated in the same way. According to Rent, a glass vessel holding the displaced water

---

[1] This omission is, itself, an operational composition. It is the inverse operation of logical multiplication. See *Proceedings of the Société de Physique et d'Histoire naturelle de Genève*, 1941, 58, p. 155.

would have to be larger than the cylinder 'because there is more water than the weight of this metal cylinder ... If it's the same size, only part of the water will go in.' It was only when the demonstrator urged him on by asking, 'What counts, the weight or the room it takes up' that Rent finally saw the light: 'The room! Oh, it's the same. The glass will have to be like the cylinder!'

If we pass on from these simple equalities to additive compositions of equivalent objects, we find that the more complex the situation becomes, the later the child applies the correct method of reasoning to it.

This is what we also found when dealing with the composition of weights (Chapter 11), so much so that we had to distinguish a Sub-stage IIIA, during which vacillation was rife, from a Sub-stage IIIB. In the present series of tests, too, though there is no such clear-cut distinction, some subjects still make a final appeal, during heterogeneous (and sometimes even during homogeneous) additive compositions to their initial 'thickness' schema. Thus Det, when asked to find the equivalent of the large cylinder III said at first that it was 'thinner' than the small cylinders, then – and this shows how confused he was – weighed it and suggested that it was equivalent to O and Pb. In other words, he had reverted to complete non-differentiation: 'Oh yes, they weigh the same so they take up the same room.' Rent argued in much the same way, and in his case, too, it was clear that the response was secondary and purely residual. Similarly Bon, who said that the greater weight of the lead might make 'a slight difference', but that this difference could be ignored because the lead behaved as if it were an aluminium cylinder, reverted to his earlier notion as soon as he was asked to compare (R + B) with (Pb + O): 'The water will rise higher (with the lead); the lead makes more pressure when it gets into the water.' Reminded that the individual terms were equivalent he saw that he had been mistaken, but then he thought about the weight and weakened again: 'Perhaps the lead will make the water rise higher because it's heavier.'

Other subjects, including Kel and, to a lesser extent, Clan, produced the correct additive compositions as soon as they had grasped the simple equivalences. How did they get over their final difficulties? Precisely as they did in the case of simple equivalences: by a new dissociation of the 'thickness' and by recomposition of the volumes thus divorced from the weight. Thus when Bon was asked to compare (CL + BR + Pb) with (R + O + B) he said, 'If we turned all these (the second set) into a ball they would be lighter but they would have the same pressure, the same height and the same size (as CL + BR + Pb), so the water would rise as high.' Even Det discarded his original prerelation when he realized that he had to go by the volume of the large cylinder: 'These three take up as much room as that one.'

As they gradually proceed to constructions, our subjects also gain increasing respect for the experimental evidence and are thus led to the elaboration of the displacement law. However, it must be stressed that though the elaboration of an inductive law and the deductive composition of concepts are two sides of a single coin, they nevertheless reflect two distinct attitudes. Thus when the child first discovers that, contrary to his predictions, the lead causes the water to rise neither more nor less than one of the aluminium cylinders, and concludes that the lead will do the same in all similar situations, he is not, *ipso facto* capable of composing $Pb = X$, $X = Y$, hence $Pb = Y$. Without the symmetrical composition of the equalities of the reversible construction of an additive set there can be no trust in the regularity of experience (as we saw at Stage I), and without that regularity there can be no composition.

In the particular case under discussion, the common denominator of these two attitudes is the dissociation of 'thickness' and the construction of new relations based on the volume alone and lending themselves to both reversible composition and to regular experimental verification. In other words, the correct reading of the experimental data not only gives rise to an empirical generalization, i.e. to a more or less plausible system of inductive analogies, but also leads to a dissociation of the initial schema and to a rational reorganization of the prevailing ideas. This is precisely what distinguishes Stage III from Stage II, during which every composition remains analogic and induction is not based on reason. But once children discover that, as Kel put it, 'it's the room that matters', they can simultaneously proceed to reversible operations (i.e. they can mentally equate the volume of a cylinder with that of the coil of clay, as Bon did) and feel certain that what they have done is precisely what happens in reality.

§4. *Stage IV : the immediate deduction of the law; compositions based on the volume alone*

Children at Stage III thus end up with the discovery and even the formulation of the displacement law. But unlike subjects at Stage IV, they start out from the global prerelation of 'thickness', and only dissociate the volume from the weight during the tests. At Stage IV, by contrast, they seize on the volume straightaway and also understand the phenomenon they are asked to explain. This difference between the two groups provides us with yet another example of the time lag between the composition of weights and that of volume.

Here, then, are some Stage IV responses, beginning with a transitional case who still appeals to the weight:

BAL (9; 3) states that B = R and R = O. 'And what if I drop in B and O?—*They'll be the same.*—And Pb and R?—*They'll be different; the lead is heavier, so the water will rise higher. Oh no; it's the same, because they take up as much room as each other.*'—(Experiment.) 'And what if I stand R up and lay B on its side?—*The same.*

'And (B + Pb) with (O + R)?—*The water will rise as high; it'll take up the same room on both sides.*—And what must I put in the water to make it rise like that (the large cylinder III)?—*You'll have to put in two.*—Which two?—*It makes no difference, they all take up the same room. Oh, no; you'll have to add one more.*—Which?—*It doesn't matter which.*

'And what about this lump of wax (W) and R?—*We'll need two pieces of wax. The wax takes up more room, it is bigger.*—Make sure!' He puts the wax in the water. '*Oh, no; it'll be the same as just one.*—And if I cut it into bits?—*It'll be the same; it'll take up the same room.*—The same room as a cylinder?—*Yes.*—What if I put in (Pb + W) over here and (R + O) over there?—*The lead will take up as much room as the red, the red as much as the orange, and the orange as much as the wax; so it'll be the same.*—And (Pb + BR + R) with (W + O + B)?—*It'll be the same, they all take up the same room.*'

DUB (9; 10). '*The water will rise as high, because they take up room inside.*—And what about B and O?—*The first will make the water rise a little more, because it takes up more room at the bottom. Oh, no; it's the same, because they're both the same size.*

'And what about Pb with R?—*The water will rise as high; they're the same size. It makes no difference that one is heavier.*—And with B?—*The same.*—And if B is lying on its side and Pb is standing up?—*The two will still be the same. It makes no difference that the lead is heavier, it still takes up the same room.*—And (B + Pb) with (R + O)?—*It's the same thing, because they're the same size. They're not the same weight but they take up the same room.*—And what about III and (R + O + B stacked horizontally)?' He looks carefully and measures them with his thumb and index finger. '*They'll be the same, or almost.*

'And this clay?—*I'll have to see.*' (Experiment.)—'And (CL + Pb) with (R + O)?—*The same. They are all the same size, because we've tried the clay with the lead and the lead is the same size as the others.*—Tell me again why the weight doesn't matter.—*It's like in the air. Aluminium spheres are lighter but they take up as much room as heavier ones.*'

LIE (12; 0). '*The water will rise* (with B).—Why?—*Because the metal takes up room.*—And what about R?—*As well, it's the same volume.*—And if R is equal to C, will B and C do the same?—*Oh, yes.*—And what about this heavy one (Pb)?—*I'm not sure.*—Is it the weight that counts, or the room it takes up?—*I think it's usually the volume, so it'll be the same.*—Why did you think that the weight could make a difference?—*Because you can never tell. I asked myself if it isn't the one rather than the other.*—And how did you decide it was the volume?—*Because if something takes the place of the water it must rise higher. So it's the space that matters, and not the weight.'*—(Experiment.)—'And with these (N + O)?—*The same, because they're the same size.*—And these (Pb + R) with (O + B)?—*The same, of course; because they're still the same volume.*

'And what about this one (CL)? I know you can't tell in advance, but what do you think?—*It makes no difference whether it's light or heavy, we have to see how much room it takes up.* (Experiment.) *It's the same.*—What if we turned the clay into a cylinder?—*It would be just the same as with the others, because it's the volume that counts. So if the clay makes the water rise as high, it must be the same size as the rest.*—And (CL + Pb + BR) with (B + O + R)?—*It's the same. Three times as much on both sides.*—And if I lay these on their sides?—*They'll take up the same room.*

'Now look. If I pour the water that came up when I put the cylinder in, you see (the demonstrator points to the difference in level), into these small (cylindrical) jars, would the jars have to be smaller, larger or the same size as the lead cylinder?—*Exactly the same, because it's the volume of the lead which made the water rise, and not its weight.*—Tell me once again why it's the volume that matters and not the weight.—*The weight can only sink down, and when it's at the bottom the water rises to wherever it can, like the air, it takes up what room it can, and when something takes up its place then it rises. It's forced up, there is as much water as before, so if something takes its place, it must rise.*—Can't it expand?—*Well, it doesn't.'*

cos (14; 0). '*The water rises because the cylinder takes up space in the water, so the water which has been pushed away takes up space on the top.*—And Pb?—*It's heavier, but I think it will do the same; the weight plays no part.*— Why not?—*Because it has to go to the bottom of the glass, and once it is there it makes no difference whether it's heavy or not.*—And (Pb + B) with (R + O)?—*The same again, twice as much on both sides. The weight makes no difference.*

'And this (CL)?—*I'll have to see.* (Experiment.) *Oh, yes, it's still the same volume.*—And if we made a cylinder out of this clay, what dimensions would it have?—*The same. It'll be taller but thinner, or if it's thicker it'll be shorter.*— And if it has the same height?—*Then it must also have the same width.*— Why?—*Because the water rises to the same height, so it has to be the same size.*

'What if I pour the water that has come up into one of these glasses?— *It'll have to be exactly the same size as the cylinders, because that's the room the cylinders took up in the water which rose to the top.'*

At Stage IV, therefore, all our subjects, except for Bal, whom we have treated as a transitional case, base their compositions on the volume right from the start. Even Lie, who wondered about the weight 'because you can never tell', quickly discarded this factor not only because 'it's usually the volume' that counts, but also because 'if something takes the place of the water it must rise higher. So it's the space that matters not the weight.' In other words the role of the volume is no longer deduced from the experiments, as it was at Stage III, but constructed by deduction. Second, all these children are immediately successful with simple as well as with additive compositions, and also produce the correct explanations. Third, and thanks to the immediate grasp of the operational and reversible character of their compositions, most of these children realize straightaway that the volume of the displaced water is equal to that of the submerged solid. Thus, while Rent still wavered when asked whether the displaced water would fill a cylindrical vessel of the same size as the metal cylinders, Lie said straightaway: 'Exactly the same, because it's the volume of the lead which made the water rise, and not its weight.' Cos, too, said promptly: 'It'll have to be exactly the same size ... because that's the room the cylinders took up in the water which rose to the top.' The 'room which rose to the top' is thus the complete solution of the problem of the displacements of volumes, albeit expressed in childish language.

§5. Conclusion: the composition of volumes and the relationship between reversible operations and experimental induction

We must still examine two final problems: that of the relationship between the composition of volumes and the compositions we have discussed in earlier chapters; and that of the relationship between operational constructions and experience in the elaboration of the displacement law. It goes without saying that these two questions are closely interrelated, and that the second takes us back to the general problems of the links between operations and experience which we left unsolved at the end of Chapter 11.

But first of all we must ask a preliminary question: can these compositions which demand a prior dissociation and which are bound up with the elaboration of a law be legitimately compared with weight compositions? We believe that they can, and this because the concept of weight also calls for a prior dissociation (between it and the quantity of matter) and a prior regrouping (as a function of the density). Even the concept of 'quantity of matter' implies such formative processes. And though the displacement law which we have used for the evaluation of volumes is more complex than that governing the behaviour of the balance, it is so only because the concept of volume is more abstract and hence more difficult to grasp than the other two. For the rest, each one of these three compositions necessarily involves the elaboration of an experimental law, and that is why the double problem we are now examining is a very general one. The reason why we have reserved the discussion of the relationship between experience and reversible compositions to the end is that this is the most convenient procedure, and also because the displacement law supplies a particularly striking example of this relationship.

By and large, the construction of volumes follows the same course as that of weights, except for the usual time lag. The starting point is the same: no compositions at all during Stage I. At Stage II the child can handle the homogeneous compositions of weights by simple equivalence or addition, but, as we saw, these are in fact compositions of quantities of matter because, in them, the weight is proportional to the substance. The composition of volumes shows a systematic time lag at this stage: it remains purely analogic or transductive. For the rest, compositions involving objects differing in weight but not in shape are successful, and so are compositions with cylinders of the same weight, in either case thanks to the dissociation of the weight from the 'size'. Hence it might seem that the logic of volumes is ahead of the compositions described in Chapter 11, but his is an illusion since 'size' in this context is nothing but the direct correlation of the volume with the quantity of matter. By contrast, all compositions involving objects of different shapes and calling for a direct grasp of volumes as such, prove abortive. At Stage III all weight compositions are successful, and it is thanks to them that the child can proceed to the logical correlation of levels. To that end, he first appeals to the weight and then disregards it thus isolating the volume in the wake of his experimental discoveries. But though this means clear progress in logical construction and may even lead to the discovery of the displacement law, it is clear that we cannot yet speak of the composition of volume as such: it is only at Stage IV that this composition finally appears.

Once again, therefore, we discover a systematic time lag between the construction of volumes and that of weights, and we now see the reason for it. Thus if we take a metal or clay cylinder and ask our subjects whether its weight is conserved when it is shifted from the vertical to the horizontal position and also whether it displaces the same amount of liquid, we shall find that before about the middle of Stage II they will deny both equivalences, only to grant them at the end of this stage. By contrast, when it comes to volumes nearly all subjects throughout Stages I and II and almost half of them at Stage III believe that the level of the water will differ according to the position of the immersed cylinders: the vertical cylinder is 'thicker', 'takes up more room', 'pushes on both sides', 'is heavier', 'is taller' (Stage II); 'is bigger', 'has more force', 'takes up more room', 'makes the water rise higher', etc. (Stage III). Quite plainly, this difference between their reactions to the weight and the volume respectively is due to the fact that, whereas the balance helps to correct the subjective weight impressions and hence suggests that the position of the object is immaterial, the grasp of volume involves a whole series of relations of which the three spatial dimensions are only the most obvious ones: the child must also realize that the submerged solid and the liquid do not suffer compression and hence take cognizance of the mass and the forces associated with the material texture of both. Hence, when he is asked whether a cylinder occupies the same space in the horizontal as it does in the vertical position, he will be thrown back on subjective impressions and seize on a dominant quality such as the height (in which case he will say that the vertical cylinder is 'bigger', etc.) or the width (in which case he will say that it is 'thinner', 'smaller', etc.). While the weight of an object thus appears as a concrete quality of matter, the volume becomes an abstraction as soon as it loses its links with the apparent quantity of matter. This explains why young children invariably lump the three together under the heading of 'thickness' and why they find it so difficult to discover an invariant capable of accounting for the 'room' a solid 'takes up' in a liquid.

Hence the time lag between the compositions of volumes and those of weights. Why, in fact, do subjects at Stage II find it so much easier to conclude that $A = A''$ if $A = A'$ and $A' = A''$ for the weight of solid bars than they do for the volume? The answer is that, in the first case, they argue about solid physical constants (they appreciate that a rigid bar will not change its weight if it is shifted around), while in the second case, e.g. the displacement of water by two pairs of two cylinders, they have to consider so many complex relations that they begin to have doubts about the regular behaviour of the immersed objects.

May we then take it that the time lag between the respective compositions of substance, weight and volume does not involve formal or logical reason as such but only its content? If we said that, we should be guilty of oversimplification; it is quite obvious that the child feels so much less certain about displacements of levels and volumes than he does about weights measured on the balance because, in the first case, the relations involved are not nearly so well composed as they are in the second and hence do not give rise to more than empirical induction.

But above all, if we look at the time lags between the compositions of substance, weight and volume respectively we cannot but conclude that the deductive or formal factors and the (perceptive and experimental) content develop as one inseparable whole. Why, in fact, does the child fail to compose weights until after he has composed quantities of matter? The reason is that weight (or the conservation of weight) presupposes the existence of matter (or the conservation of matter), but not *vice versa*. Similarly, compositions of volumes lag behind those of weights because their conservation depends on the consistent behaviour of matter, and according to the child, the latter, i.e. the incompressibility or 'hardness' of objects, rests on the conservation of weight (see Chapter 6). Now is this implication — of matter in weight and of weight in physical volume, but not *vice versa* — the result of an operational construction, or does it spring from the experimental findings? It is clear once again that these two factors are interdependent: the objective discoveries depend on a system of operationally grouped concepts and the latter are nothing but possible actions.

This leads us, or rather brings us back, to the problem of the relationship between form and content, or more precisely between reversible composition and experimental induction.

Let us try first of all to define the characteristic feature of the composition of volumes discussed in this chapter. The starting point is a lack of differentiation of the basic concepts, i.e. the belief that all three relations (substance, weight and volume) are directly proportional to one another. In other words, A is taller than B, or $(A \xleftarrow{1} B) = A$ is bigger than B, or $(A \xleftarrow{2} B) = A$ is thicker than B, or $(A \xleftarrow{3} B) = A$ is heavier than B, or $(A \xleftarrow{4} B) = A$ is stronger than B, or $(A \xleftarrow{5} B) = A$ makes the water rise higher than B, or $(A \xleftarrow{6} B)$. Hence $(A \xleftarrow{1} B) \times (A \xleftarrow{2} B) \times (A \xleftarrow{3} B) \times (A \xleftarrow{4} B) \times (A \xleftarrow{5} B) = (A \xleftarrow{6} B)$. At Stage I, therefore, there is no operational composition whatsoever: the child believes implicitly, and regardless of the absurdities to which this assumption leads him, that all three relations are directly proportional, and it is this belief that we tried to express symbolically so as to illustrate the conditions under which the subsequent grouping is effected.

Now the factual data we present to the child (and the beginning of the test has the express and sole purpose of familiarizing him with these data) are such that they cannot be treated in this way. Thus when the child is asked to compare the lead cylinder A with the aluminium cylinder B, he finds that $A = B$ in respect of relations 1, 2, 3 and 6, but $(A \xleftarrow{4} B)$ and $(A \xleftarrow{5} B)$ when he equates the force with the weight. Similarly, the clay coil gives $(A \xleftarrow{1} B)$, $(A \xleftarrow{2} B)$, $(A \xleftarrow{3} B)$, $(A \xleftarrow{4} B)$, $(A \xleftarrow{5} B)$, and $(A \xleftarrow{6} B)$, etc. Now what happens when the various constructions based on the principle of proportionality prove to be contradictory? At Stage I, the child prefers his global schema to operational coherence; moreover, three factors militate against composition, and, incidentally, help us to a clearer understanding of the subsequent construction. In the first place, the child will sometimes behave as if the spatial dimensions were proportional to the weight, and at other times as if the converse were true, whence he is faced with a permanent contradiction. In the second place, when one of the relations is changed abruptly (for instance, when the recumbent cylinder is stood on end) the child considers the change in isolation, i.e. he fails to compose it with the inverse change (the height increases but the width diminishes proportionately). In the third place, because he sticks to the belief that the relations are globally fixed—a belief that is constantly refuted by the real transformations—he begins to have doubts about the consistency of experience and also about his original hypotheses. It follows that the success of composition depends on three factors: (1) dissociation of the qualities involved, i.e. division of classes or abstraction[1] of the qualities of weight and force so that only the spatial dimensions are retained (provided, of course, the solid is heavy enough to be completely submerged); (2) multiplication of the remaining relations with due compensation for all changes, i.e. a system in which the volume of a solid can be expressed either qualitatively or quantitatively in terms of the dimensions and hence be correlated with that of the displaced water (see the responses of Lie and Cos; Stage I·V); (3) a set of logical and experimental constants.

We can now determine the respective roles of experience or experimental induction and of composition, i.e. of operations as such, in the logical construction of the displacement law. It is clear, first of all, that the three conditions we have just mentioned cannot be satisfied without experience: experience alone supplies the contents of an operation, i.e. decides which operations must be effected and in what sense. Why does the level of the water depend on the volume and not on the weight of the solid when the latter is heavy enough to be fully submerged? Why is there a rise in the level, rather than a compression, when a heavy metal cylinder is immersed? Why are the weights and the volumes of solids not necessarily proportional to each other? Why does the volume of a solid remain constant, i.e. why is every change in one of its dimensions compensated by changes in another dimension? In brief, it is experience alone which can tell what relations must be dissociated or composed, how they must be composed, and what real or physical constants must be taken into account.

But it is equally clear that none of these experimental data can be read off from simple observations or experiments, or even conceived in the absence of formal compositions. Let us look at the

---

[1] In the operational sense, abstraction is the inverse operation of a logical multiplication. See *Proceedings of the Société de Physique et d'Histoire naturelle de Genève*, 1941, 58, p. 155.

third condition first: despite the regular course of the universe and the existence of physical constants, we all know that experiences do not repeat themselves in every intimate detail and that the mechanisms we are concerned with involve a large measure of abstraction. Thus if the lead cylinder A causes the water to rise as high as the cylinder A' and if A' = A" why are we entitled to conclude that A = A"? It is not exaggerated to say that our only justification is logical necessity,[1] because if we were to follow the destiny of every drop of water and of every molecule of the metal, we should find that a great many changes take place between A and A". And why should the effect A remain equivalent to the effect A' when we repeat the experiment? Once again, simply because the composition demands that A = A and that A' = A'. This means that, if we should discover from an experiment conducted with a high enough degree of precision that A is no longer equal to A" or A', we would not say that A, A' or A" have ceased to be identical to themselves and hence 'unthinkable', but that they have changed physically, which does not prevent us from obtaining their original state by abstraction and hence from asserting that if A = A', and A' = A" then A = A", now as ever. In other words, the experimental content suggests what operations are needed but, as the physical world is transformed and flows on, reason immobilizes the flow and provides the means of retrieving what states it requires by rendering the world reversible in thought. Next, it does the same with new observational data and so on, thus constantly correlating the present with the past. Hence, though the lead or the other cylinders do undergo minor changes during the experiments, and though the water evaporates or expands and the vessels themselves change their shape slightly, we can nevertheless take it that A = A and that A = A' = A" by reconciling the new data with the old with the help of other relations. It follows that, while the child has not yet learned to deduce A = A", he can have no confidence in experience, and *vice versa* : experimental and operational reality (no matter whether they are based on physical or purely logical operations) are constructed simultaneously.

Similarly, in respect of the first and second conditions: it is impossible to dissociate the volume from the weight (first condition) or to co-ordinate the complex relationships involved in the concept of physical volume (second condition) without recourse to a reversible mechanism, i.e. without logical groupings or arithmetical and geometrical groups. Much as the laboratory physicist can understand nothing, record nothing and observe nothing without the mathematical apparatus that serves him as an instrument of

[1] A. Lalande, *Les Problèmes de l'induction et de l'expérimentation*, Paris, Boivin.

explication at the end of his analyses and as a language right from the start, or indeed as an instrument of perception (since, as H. Poincaré and P. Duhem have said long ago, the physicist cannot 'see' electricity except in the oscillation of a needle across a scale, or any other phenomenon except through a set of similar abstract signs), so the human mind from its earliest beginnings cannot grasp the precise meaning of such concepts as 'quantity of matter', 'weight' or 'volume' except through seriations, equivalences, abstractions (class divisions), multiplications of relations, etc., i.e. through formal, operational groupings, of which numerical or metric quantification is the direct result. For example, before he discovers that the level of the water depends on the volume, not the weight, of an object (heavy enough to be completely submerged) – a dissociation dictated by the most elementary observations – the child has to pass through no less than four stages, covering the first four to eleven years of his life, and involving the successive operational constructions of all three invariants. The realization that a lead cylinder of the same dimensions as an aluminium cylinder will displace the same volume of water thus depends on the construction of a host of logical groupings (equivalences, seriations and multiplications of asymmetrical relations). Similarly, the realization that a horizontal cylinder occupies as much space as a vertical cylinder of the same dimensions implies the whole of logic. And to appreciate the simplest empirical significance of the fact that a coil of clay will produce the same change in level as a metal cylinder calls for reversible operations with the help of which the subject can then go on to construct a clay cylinder of the same dimensions as the others, and even come to realize that he can pour the displaced water into a cylindrical vessel of the same volume.

No long commentaries are needed to show that the same compositions, but in the parallel form of physical operations – the reader will recall the parallelism we established at the end of Chapter 11 – also led the child to the conservation of volume (Chapter 3), to atomism which helped him to extend that conservation to the case of dissolved substances (Chapter 6) and to the schema of compression and decompression at constant corpuscular volume (Chapters 7–9). In respect of the first point, it is quite obvious that the conservation of volume, no less than that of substance and weight, depends on the composition of equivalences and also on additive compositions, and that the physical invariants thus constructed constitute a victory of operational reversibility over the flow of events. As for atomistic compositions, we saw quite clearly that they are a simple extension of the first on the corpuscular plane: because they transcend the limits of perception, they call all the more urgently for a deductive mechanism capable of reconstructing the real facts. In particular, there is a remarkable parallelism between the operational or formal and the experimental or inductive approach to the conservation of the dissolved sugar (nascent atomism) (Chapters 5 and 6), and the tests discussed in this chapter. At Stage I lack of conservation is coupled to a total indifference to experience; at Stage II the nascent atomism and the conservation of substance go hand in hand with incipient submission to the dictates of experience, but there is still a lack of co-ordination; at Stage III atomism and the correct composition of weight is associated with the development of systematic induction and complete submission to experience; Stage IV, which sees the complete composition of matter, weight and volume goes hand in hand with complete deduction from experience. Hence there is a complete correspondence not only between the compositions we are dealing with here and those at work in atomism, but also between the inductive approach to both, from the initial level of total indifference to experience up to the final level on which the composition of equivalences and additions help the operational mechanism to carry the day with extraordinary deductive force.

We are therefore entitled to conclude that, in all the fields we have examined in this book, the content of thought is complementary to the form, the content consisting of the world such as the child perceives it and the form providing him with the sole means of reverting from the state T of this world to the state $T - 1$, i.e., of reversing reality in thought. It would, however, be quite wrong to claim that the content reduces to the terms on which the operations are brought to bear (A, A', A'', etc.) and the formal structure to the reversible operations themselves ($=$; $+$; $-$; $\times$; etc.), since the terms as such cannot possibly be isolated and defined outside the relations in which they appear as so many results of the operations linking them; conversely these operations are rooted in reality because it is reality alone that imposes the various combinations or dissociations between the terms. In a sense, the mind does nothing more, therefore, than extend reality, but it does so by adding reversibility.

Returning now to the relationship between experimental induction, whose function it is to follow reality in all its ups and downs, reversible or otherwise, and deductive composition in either its logico-arithmetical form, in which abstraction is made of time and space, or in the form of physical operations whose role it is to remould the universe on the basis of reversible constructions, we find that induction, too, constitutes a form of composition—we have just seen why—but an incomplete one, i.e. 'a grouping' or a group of operations not yet closed on themselves. Hence it is quite easy to define the mixture of experimental induction and pure deduction which we have encountered both in the construction of the concepts of conservation and especially of atomism, and also in the composition and elaboration of the displacement law examined in this chapter. The construction of an experimental law is nothing but the gradual perfection of compositions, a process that is crowned with success when the relations involved are simple enough to be grouped, but which remains bogged down in induction when the compositions cannot be fitted into a reversible system, in which case the relations observed between the phenomena can only be stated but not deduced. In other words, induction is the application of reversible operations to a content that has remained irreversible, either because it consists of data that are not sufficiently known or elaborated to give rise to a coherent grouping, or else because it is irreversible in fact. In either case, the operations themselves do not culminate in a complete composition but merely prepare the way for it: whenever the induction of a law proves successful, induction eventually becomes fused with deduction.

Conclusion

We should like to conclude this study with a few remarks about quantification, logic and experience in general.

I

It would be absurd to speak of a radical opposition between quantity and quality, as if mathematics were a purely quantitative discipline and logic a purely qualitative one. As the gradual quantification of physical qualities discussed in this book shows only too clearly, the two are inseparable.

Thus every qualitative relation involves a quantity. We never perceive or conceive of absolute or isolated qualities, and always relate qualities to one another; and when we believe the contrary we are labouring under an illusion. Thus Köhler has shown that an animal trained to select B rather than A will also select C rather than B, if it is shown B and C simultaneously and if these terms (colour, size, etc.) are related in the same way as A was to B. And since qualities are interrelated, they cannot be called more primitive than quantities; the two are distinct but inseparable concepts.

However, there are three possible types of relations between qualified objects and consequently three types of quantities, each of which calls for separate psychological and axiomatic treatment.

There are first of all intensive quantities (to use Kant's classical term), which simply define the relationship of the part to the whole and state that the whole is greater than any of its parts and that any part has the same size as itself. This type of quantity is the only one to intervene in logic, but it intervenes in every logical grouping. Thus if all the A and all the A′ are B, and all the B are A or A′; and if all the B and all the B′ are C and all the C are B or B′, then we know that $A < B$ and $A' < B$; $B < C$ and $B' < C$, etc., but we know nothing about the relationship of A to A′; B to B′; A to B′; A′ to B′, etc. Similarly, if a is the difference between O and α in a series of asymmetrical relationships; a' the difference between α and β; b' the

269

difference between β and γ, etc.; then, $a + a' = b$; $b + b' = c$ etc., we know that $a < b$, $a' < b$, $b < c$, and $b' < c$, but we know nothing at all about the relationship between a, a' and b'. This is equally true if a, b, c, etc. are symmetrically colligated, since the only logical terms expressing the colligation of classes are the quantities 'one', 'some', 'any' and 'all', and since in such qualitative seriations as 'the wine A is not as good as the wine B, and the wine B is not as good as the wine C', we only know that there is a greater difference between A and C than between A and B or between B and C, but we do not know precisely how great these differences are.

Let us now suppose that, instead of simply posing $A + A' = B$ or $a + a' = b$ we could put $A = A'$ or $a = a'$. We should then have $B = 2A$, $C = 3A$, etc. (where $b = 2a$; $c = 3a$; etc.), i.e. a series of numbers or segments; whence we obtain a second type of (metrical or numerical) quantity based on the construction of units (A or a).

There is yet another possibility. Even without equalizing A, A', B', etc. or a, a', b', etc., we can establish such relations of difference between them as obey some law of construction, for instance a series of increasing or decreasing differences, of proportions, of harmonic relations, etc. The result is a quantification that, though not metrical, nevertheless goes beyond simple logic. The whole of qualitative geometry[1] is based on this third type of quantity, which we shall call *extensive*, a general concept of which the second type of quantity is but a particular case.

II

Because intensive quantities are characteristic of logic, it might seem that they result from the most elementary perceptive or intuitive relationships. However, though quantities are involved at the most primitive levels of mental activity, they are so only in a 'raw' form, i.e. they remain undifferentiated (mixture of extensive and intensive), and hence full of contradictions, except in the fleeting realm of direct perception or intuition. We might therefore expect to find that intensive quantities cannot assume a stable and differentiated form until logic itself has been organized. Now, logic is not something innate: every one of the tests we have described in this book shows that, at a certain stage of mental development, the merest change in the arrangement of the parts is enough to persuade the

---

[1] Especially the theory of the continuum (except for Archimedes's axiom, which is metric); Cantor's postulate, the theorem of Weierstrass, etc. When Kowalewski defined the 'neighbourhood of a convergent point' as containing 'nearly all' the points of the set, this 'nearly all' (= all except a finite number), though not being metric, undoubtedly went beyond the 'some' and 'all' quantifications of logic.

child that their sum (or logical addition) is no longer equal to the whole. Again, if A = B and B = C, children between the ages of seven and nine years do not necessarily conclude that A = C. Moreover, seven- to eight-year-olds are unable to seriate three weights (A < B < C), which explains why they cannot produce systematic quantifications of even the intensive type: all such quantifications are correlations of the parts with the whole, and without logic there can be no stable whole. But, as we saw, no logic means no conservation, and without conservation there can be none but the 'raw' quantities of perceptive intuition.

Whence our fundamental problem: how does the child arrive at logic in general, or at particular logical solutions? We saw that he does not do so by gradual improvements of his intuitive or perceptive methods, because perception is rigid and irreversible and because the co-ordination of reversible transformations demands a break with the perceptive structures and calls for the construction of a system of pure operations.

Now it is in this field that the present study not only validates a hypothesis we first put forward in *The Child's Conception of Number* but also proves its generality: logical operations do not appear as so many judgments out of context, i.e. as the kinds of judgment described in classical textbooks of logic, but in the form of general systems. There cannot, in fact, be such a thing as an isolated operation, for the simple reason that every operation is a virtual action, but one that can be co-ordinated with others (composition) and that proceeds in two directions (reversibility). The concrete and natural feature of operations is therefore a grouping: there can be no operations prior to a grouping because it is only thanks to the grouping that actions become operational or that operations can be constructed as such.

What, in fact, have we been able to gather from the child's discovery of the solutions to every one of the problems posed in this volume? That he proceeds to the simultaneous construction and co-ordination of the following four mechanisms: (1) Direct operations, which are constructed by any actions whatsoever provided only that the composition of any two results in an action of the same type and that the inverse action remains part of the system. Examples are, first, all actions leading to the combination of the parts into the whole (or to the separation of the parts of that whole): if A is one part and A' another, we have $A + A' = B'$; if B is combined with B' we have $B + B' = C$, etc.; second, all actions serving to fit the elements into a series of increasing or decreasing differences (or of displacing them by translation, rotation, compression, etc.): if a is the first difference and we add a' to it, we have

271

$a + a' = b$; $b + b' = c$, etc.; and finally, actions serving to combine two such series: A is both longer and thinner than B, etc. (2) Inverse operations: the above actions do not become operational until they are correlated with their inverse: separation with combination, displacement with placement (or replacement with displacement, in which case the inverse of the inverse becomes the direct operation), increasing differences with decreasing differences, etc. In fact, everything we have learned about the origins of reversibility (Chapters 1–6, etc.) tends to show that it is the *sine qua non* of the construction of operational systems; conversely true reversibility (in contrast to the 'empirical return to the starting point') cannot be properly constructed unless it is based on the entire set of operations. In other words, direct operations are necessarily based on their inverse and *vice versa*. (3) But before the mind can grasp this mobile equilibrium, another condition has to be satisfied, and observation has shown us once again that it is synchronous with the last: that, if a given conclusion is reached by two different paths it is recognized as being 'the same'. This is what we refer to by associativity. Thus if we divide the ball C into the parts A, A' and B' and then recombine A + A' into B and then add B', or, omitting A, combine A' with B', the child who has arrived at the logical stage will always grant that $(A + A') + B' = A + (A' + B')$, when previously he did not necessarily do so. (4) If we now compose a direct operation with its inverse (for example, by flattening a ball into a disc and then rounding the latter into a ball), it looks as if we had done nothing at all. This is what we mean by *identical operations* which guarantee the identity of the whole or the part. Now, as we saw time and again, the simple identification invoked by our subjects ('you took nothing away and added nothing') only appears in conjunction with reversibility and the grouping as a whole.

All these processes appear simultaneously as soon as the child begins to think of the transformation as such, instead of clinging to each perceptive form as if it were absolute and self-sufficient. The repercussions on quantification which we have noted and which also provide the best test of the completion of groupings are the *a priori* affirmation of the conservation and the construction of atomistic schemata. To determine the gradual construction of a grouping of operations we need merely look at the spontaneous answers offered by our subjects, because once they have passed beyond the simple intuitive stage, they invariably and automatically invoke direct operations, reversibility and identity, and also because all their statements reflect their grasp of associativity. However, their groupings might well have remained purely rudimentary, were it not for their advance from empirical and probability statements to deductive and constructive arguments, thanks to which the conservation is affirmed as a logical necessity and justified by the construction of a system of atomistic relationships.

III

Before demonstrating how completed groupings engender extensive quantifications as soon as the intensive quantities defined by the relationship of the parts to the whole are constructed, we must first stress that the operations whose groupings we have been describing can be of two types. When supposedly constant objects are combined into classes or seriated according to their relations, the resulting classes and series are independent of time and space, in much the same way as the numbers or arithmetical equivalences that can be established between them. We speak of logico-arithmetical operations when the following two conditions are satisfied: (1) the elements on which the operations are brought to bear are individual and constant objects, and (2) the operations are not affected by spatio-temporal conditions.

Now, when our subjects combine (in fact or in thought) bits of the clay into a single ball or dissolved sugar grains into a single lump they compose the relationship of the part to the whole in the same way as they would combine individual elements into classes or into relations between sub-classes and total classes; however, in the present case the whole, i.e. the ball or the lump, is not a class but an individual object. It might be objected that the difference between the relations of the part to the whole or of objects to the entire class concerns the representation or the content of thought rather than logic, but this view is quite wrong. We can easily make a purely formal distinction between transitive inclusions of the parts in a whole (e.g. Socrates's nose is part of his face; Socrates's face is part of Socrates; therefore his nose is part of Socrates, etc.) and equally transitive inclusions of individuals in a class or of classes in classes (e.g. Socrates is an Athenian; Athenians are Greeks; therefore Socrates is a Greek), because the first are not transitive with respect to the second: Socrates is an Athenian, a Greek, etc., but neither his head nor his nose are Greeks or Athenians, although they are parts of himself; and if he had lost his nose he would still be an Athenian, or a Greek, etc., but he himself would no longer constitute the same whole. We can thus contrast partition or partitive addition (the combination or division of parts) with inclusion or logical addition (the combination or exclusion of objects *qua* elements of classes) by distinguishing the first of these operations from the second on the basis of the following two criteria:

(1) they are infra-logical, i.e. the 'whole' constituting their upper limits is the individual object (even if it is as large as the universe); (2) they treat the parts and the whole as spatial or temporal elements because they delimit them by divisions, not by abstract distinctions.

Similarly, when our subjects treat the successive states of a ball, a lump of sugar or a maize seed as results of expansions or contractions, decompressions or compressions, etc., it is clear that they are composing these relations in the same way as they would asymmetrical relations open to additive or multiplicative seriation. However, what is involved here is no longer the seriation of objects as such, or the preservation of constant relations between them, but a seriation of the successive states of one and the same object. Now, the relations between these states are not random but consist of spatio-temporal placements whose transformations are displacements; and much as partitions help us to compose objects that can later be classified, so placements and displacements are infra-logical operations engendering relations that can later be seriated.

In brief, spatio-temporal or physical (i.e. infra-logical) operations have the same formal structure as logical operations, but quite a different operational significance, and that is why the two cannot be composed, even though they are organized with the help of the same types of grouping. Like logical operations, physical operations can thus appear in qualitative form, with a purely intensive quantification and be quantified extensively or metrically.

IV

The transition from intensive to extensive quantities is a metric one, and this in the double sense of logical and spatio-temporal operations, because it proceeds, as we saw time and again, not by the addition of new operations but by the simple regrouping of the operations involved in the previous grouping.

In fact, in the entire field we have been exploring, all developments have proved to be synchronous: contrary to what we might have expected, there is no primary stage of intensive quantification followed by a later stage of extensive or metric quantification. What is given in the first place is something quite different: a 'raw' quantity of the intuitive or perceptive type, both undifferentiated (i.e. intensive as well as extensive) and incoherent. Next comes the construction of intensive quantities with the help of logical and physical operations whose development we can follow step by step; and as soon as this first type of quantification is completed, metric or extensive quantities in general appear simultaneously, as it were, out of the blue. This paradoxical situation reveals the existence of a genetic mechanism that, in turn, throws a great deal of light on the relationship between logic and the mathematization of qualities.

We are not, of course, referring to the child's usage of such acquired notions as prime numbers, but only to the concrete and spontaneous measurements elicited, for instance, by the experiments described in Chapter 9.

As far as the quantity of matter or the substance, i.e. the simplest form of quantity, is concerned, we have not (except for the homogeneous bars described in Chapter 11) conducted new experiments into extensive or metric quantification, because we have described the results elsewhere.[1] As the reader may recall, it is as soon as the child starts to assume the conservation of liquids decanted from one vessel into another and hence ends up with what is, at least, an intensive quantification, that he learns to solve the first problems in metric or extensive quantification. Thus if the content of a vessel B is poured into the vessels $A_1 + A_2$, he will predict that if A_1 is poured back into B it will only rise to half the initial height, etc., or if B is poured into L, a narrower but taller vessel, he will take the inverse proportions into account and equalize the differences, etc., whereas previously he failed to grasp these simple metric relationships.

Now, interestingly enough, the same interdependence between intensive or logical and extensive or metric quantification also holds for the case of weights but with a time lag of one stage (Stage III instead of Stage II). Thus, at the same stage that he grasps the conservation of weight, he also grasps that, having fashioned a clay ball equal in weight to a cork, he need only cut his own ball in half to copy the weight of half a cork, etc. Now, the reader will remember how many absurd solutions he offered before he arrived at this conclusion. Again, at Stage III, he not only realizes that the two brass bars A and B are equal in weight and that if B has the same weight as the lead bar C, then A = C (intensive quantification), and also that if C = D then A + B = C + D, or A + C = B + D, etc., and more generally A + A = 2A (metric quantification). As for non-metric quantifications, our analysis of the discovery of conservation in general and of the four methods leading to it in particular (Chapter 1) has shown clearly that the invariance of the proportions is grasped at the same time as the equality of the differences and the identity of the logical elements.

Let us also recall that at Stage IV the discovery of the conservation of physical volumes involves the same synchronisms: logical

[1] J. Piaget and A. Szeminska, *The Child's Conception of Number*, Routledge & Kegan Paul, 1952, Chapter I and especially Chapter X.

grouping entails metric compositions (Chapter 12) as well as non-metric, extensive compositions (proportions).

The spontaneous development of atomism proved the generality of this process: born of infra-logical partitions and displacements, atomism is the prototype of intensive quantifications: its sole aim is to provide an explanation of conservation. To that end, the 'grains' are equalized and reduced to a set of units, this type of quantification going hand in hand with an implicit metric system: the grains constitute a denumerable set whose elements can be put into one-to-one correspondence with the positive integers.

Now, the reason why the psychological constructions of intensive and extensive metric quantities are synchronous is undoubtedly that the underlying operational mechanisms are so closely interrelated. In fact, we cannot point to any special operation leading to the construction of extensive metric quantities; what is new, therefore, is the synthesis, not the operational form. True, we have the iteration $A + A = 2A$, without which there could be no metric system and which distinguishes numerical relation from the logical addition $A + A' = B$, or from the tautological identity $A + A = A$. But iteration is only the addition of the unit A to itself, and that unit merely results from the generalization of what substitutions can be effected with A, A', B', etc. in the additive grouping $A + A' = B$; $B + B' = C$, etc. What distinguishes mathematics from logic, i.e. extensive from intensive quantification, is therefore the transition from the grouping to the group; the latter is a new operational organization but without the addition of any special operations that differ radically from the logical.

We have dealt with this transition in our discussion of the composition of weights (Chapter 10, §3), and we must now show that the explanation we offered is generally valid, i.e. that logical groupings are transformed into mathematical groups by an operational synthesis of the grouping of classes (equivalences) and the grouping of asymmetrical relations (differences) through the process of equalization of differences.[1] In fact, when a set of elements such as a piece of lead A_1, or a set of three bars A_2, A_2' and B_2', is defined by its qualities alone we can group the elements: (1) by seriating their differences: thus $0 \xrightarrow{a} A_2 \xrightarrow{a} A_1$ would mean that A_2 is denser than zero and that A_1 is denser than A_2, etc.; (2) by combining them into classes: thus the lead will form a special class A_1, and the three bars another class ($C_2 = A_2 + A_2' + B_2'$), because A_2, A_2' and B_2' are equivalent in respect of all their qualities and because A_1 has other qualitative characteristics. We might also discover that these

[1] See J. Piaget and A. Szeminska, op. cit., pp. 240–3, and the Proceedings of the Société de Physique et d'Histoire naturelle de Genève, 1941, 58, p. 123 (Theorem VII).

objects have the same weight, in which case we can construct the class D ($= A_2 + A'_2 + B'_2 + A_1$, which has now become C'_2) containing the entire set of the same objects but without excluding the particular qualities by which they are distinguished. (3) Let us now suppose that we disregard the special qualities and retain only the equivalences; then every one of the terms is raised to the rank of a unit A and the class D is immediately defined by $D = A + A + A + A$. But in that case how can we distinguish the As from one another? Once again by seriation. Now the logical groupings (1 and 2) do not help us to combine the classification of equivalences and the seriation of differences into one operational whole, because we cannot construct a qualitative series while reserving the right of permuting the terms (i.e. of considering them as equivalent). The only seriation that allows us to distinguish between the equivalent units A will therefore be one that disregards their particular qualities and only bears on their differences in order, for in that case, no matter what order we chose we shall always have $A + A + A + A = D$. It is in this sense that the numerical group (3) can be said to result from the synthesis of logical groupings of equivalences (2) and of differences (1) by the generalization of both.

It should also be stressed that the same mechanism is at work in spatio-temporal physical operations. We can, in fact, treat our objects (the bars and the lead) either as elements of classes or else as parts of one and the same whole (= the total weight resulting from the addition of the partial weights). In that case the metric system becomes a synthesis of partitions (once the parts can be substituted for one another in the form $A + A + A + A$) and placements (each part can be distinguished by the place it occupies in the spatial arrangement). Thus the child constructs his atomistic models by means of the partition of matter (division into grains) and of groupings of placements or displacements (simple displacements or compression and decompression, etc.). Now, as soon as that qualitative construction has been completed, it engenders virtual measurements because the subject treats every grain as a unit equivalent to all the rest. But, how does he distinguish them from one another, seeing that he asserts the conservation of the total 'number' of these particles, none of which he can perceive, in the sugar water or in the maize flour? He does so by 'placing' them beside one another or beneath one another in his mind, so that no two ever occupy the same spot simultaneously.[1] A measurable quantity is therefore the synthesis of a partition and a placement (or displacement) by equalization of the unit-parts or by generalization

[1] The total arrangement remains the same if any two elements are changed round; the order is thus vicarious, as it is also in finite numerical ordination.

of the idea of order (all rows that can be constructed by putting the same units end to end have the same value), much as number is the synthesis of a class and an asymmetrical relation by the generalization of equivalences and seriable differences.

As for extensive, non-measurable quantities, it goes without saying that they, too, result from a comparison of the different parts of one and the same whole to one another, not to the whole, but what we have here is the equalization of ratios (proportions, etc.), and no longer of units or of differences as such.

V

Which, then, are the precise roles of experience and mental activity in this overall construction embracing the inclusion of classes, the seriation of relations, number, partition, placement and measurement? We have met this problem more than once in the course of this work, and quite particularly when looking at the relationship between the experimental content of operations and their deductive structure.

To begin with, we came to distinguish two factors, namely the relationship between physical and logico-arithmetical operations, and the relationship between both and experimental induction. Now, it would be wrong to consider the first of these operations as the content of the second or as being more closely bound up with experience. We saw that both are constructed simultaneously and that they spring from a common source, the first helping to structure the parts of the object and their placements, and the second to combine objects or their relations. Each is therefore as formal or as experimental as the other.

What, then, is the common source of these two types of operation? This is the crux of the problem of deductive form versus experimental content, or of composition versus induction. Now the common source of operations in general is action in its most concrete and sensori-motor form. Every operation, as we saw time and again, is a reversible action. Irreversible actions are not operations because they may lead to any result whatsoever, and this in a purely empirical way. Hence, they may result in fruitful, but never in regular, constructions. Regular constructions only begin when the actions are grouped into a closed operational system, and into one that is both associative and reversible (see II). Induction is nothing but construction in the process of grouping, and deduction begins as soon as that construction is effected within completed groupings or groups.

But though this solution is adequate to the sphere of reasoning, the problem of the respective roles of the form resulting from intellectual activity and of the content imposed by experience as such also occurs in the spheres of induction and of pre-operational mental processes. Induction is ungrouped reasoning, which may go hand in hand with true deduction; but before either emerges, there is intuitive intelligence, sensori-motor intelligence, habit and perception, and in each sphere the problem can be posed in new terms.

In general, we can say that from the lowest stage of mental activity onwards there are two factors whose relationship at first overshadows the problem with which we are here concerned. These factors are motivity, the source of all future operations, and tangible matter, the point of contact between the child and the properties of the external medium. Now, two series of investigations enable us to assert that however high we climb up the mental scale, we shall find that these factors are interdependent. Studies[1] of perception have shown that, though tangible matter provides motivity with its reference points and signals, the latter provide it with its structures. The result is an endless circle (the so-called *Gestaltkreis*). Moreover, in our own studies of perception, and above all of the sensori-motor intelligence characteristic of the first few months of mental development, we, too, were able to show that both factors are interdependent from the outset, i.e. that they are part and parcel of sensori-motor schemata in which sense data remain meaningless unless they are assimilated to repeated actions, and in which the latter fail unless they are accommodated to the perception of successive data. It is to this egocentric assimilation of reality to motivity and to this phenomenalistic accommodation of the action to the external data that we must look for the earliest contacts and the first conflicts between mental activity and experience.

Now, the whole history of groupings is one of gradual decentration, i.e. of the rejection of egocentrism in favour of closed and regular compositions, and this because the initial actions have become reversible and operational or, if you like, because the child's own actions have been incorporated in an overall system of possible actions. Not surprisingly, therefore, the history of experimental science is one of the gradual correction of phenomenalism, thanks to which all the perceived data can, sooner or later, be treated as consequences of possible operations. It is only when this happens that the original assimilation and accommodation, which remained undifferentiated and antagonistic while assimilation was still egocentric and phenomenalistic, become dissociated and complementary, the one by a grouping of operational actions based on inner necessity, the other by the application of the most probable model suggested by that grouping to irreversible processes.

[1] By the disciples of von Weizsäcker, Auersperg, *et al.*

Index

assimilation: perceptive, 211; sensori-motor, 205

atomism: beginning of, 69–9, 90, 94–5; and compositions, 81, 103; and compression, 96, 148–9; defined, 132; development of, 107–16, 120, 126–8, 177, 276–8; also, 52–53, 55, 78–80; *see also* conservation; egocentrism

Auersperg, 279*n*

Bachelard, G., 68, 78*n*
Berger, C. A., 81*n*
Binet, A. and Simon, T., 195 and *n*

Cantor, 270*n*

compositions: additive, 240, 248, 256; and atomism, 81, 103; of density, 177; deductive, 249; of equivalences, 219, 247, 253–4; heterogeneous, 204, 212–14, 220, 234, 261; homogeneous, 204, 207–23, 233–4, 238, 240, 248–9, 256, 260, 267; intuitive, 203*n*, 208, 212, 220; irreversibility of, 179; logical, 234, 238, 241–2, 248, 254, 264–5, 267; logico-arithmetical, 238–43, 248, 250; operational, 89; of parts, 132, 142–3, 168, 170; physical, 233, 243, 260, 267; quantification of, 103; of relations, 102; reversibility of, 39, 54, 80, 131–41, 220–21, 257, 260, 263; and seriation, 205, 208, 213, 222, 225; spatio-temporal, 94; of substance, 126, 146, 152, 177, 262–3; of volume, 129, 135, 171, 233, 239, 248, 260–

8; of weight, 126, 146, 152, 171, 203–31, 250, 261–5, 275; *see also* egocentrism; stages

compression, and decompression: and atomism, 96, 148–9; and conservation, 55, 105, 108, 119–24, 127–42; defined, 145; and density, 52, 145–77; *see also* egocentrism; stages

conservation: and atomism, 52–3, 55, 67–8, 78–81, 90, 94, 103, 107–18; and compression, 55, 105, 108, 119–24, 127–42; of continuous quantities, 14–15; of density, 177; experiments in, 5–7, 9–10, 12–13, 22–33, 35–8, 43, 56; and intuition, 12, 29, 60; mathematical, 19–20; in physical transformation, 69*n*; quantification of, 26–7, 38, 44–5, 54, 81, 95, 96, 102–3, 113–15; reversibility in, 10–12, 17, 39–41, 44, 54–5, 57; *see also* egocentrism; stages

continuous quantities, conservation of, 14–15

co-ordination: of equivalences, 203, 208, 217–19, 221; objective, 87–9; of operations, 17; of relations, 20, 41, 51–2, 55, 76, 114–15; in seriation, 192, 194; of substance, weight and volume, 60–3, 74, 88–9, 111–14, 130–5, 140–68, 174, 177, 211, 214, 233–8

correspondence, numerical, 172

decompression, *see* compression
density: comprehension of, 122–3;

and compression experiments, 52, 94–5, 119–24, 127–42, 145–77; concept of, 52; conservation of, 177; and intuition, 143, 151; quantification of, 148; and substance, 139, 142–3, 151–3; and volume, 137–43, 148–53, 155; and weight, 137, 139, 142–43, 151–2, 170; *see also* egocentrism

displacement, of solids in water 5, 232, experiments in, 47–56, 232–7, 243–5, 250–3; *also* 5, 232–68

displacement law, 249–50, 253, 257, 260

dissociation: of self from data, 33, 62–3; of weight from length, 176; of weight from quantity of matter, 121, 124–5, 137, 140, 152–5, 160–4, 168; of weight from volume, 137, 140, 152–5, 168, 233–4, 254, 257, 264–6

Duhem, P., 266

egocentrism, phenomenalistic, 279; in composition experiments, 213, 217–20, 227, 241; in compression and density experiments, 121, 134, 140–4, 158, 161, 166–7, 173; in conservation experiments, 29, 32–3, 38, 42, 45–6, 53, 59; in conservation and atomism experiments, 75–80, 86, 96, 101–2, 112–15; defined, 87, 96; in seriation, 200–1

equalization: of differences, 19–20, 26–30, 38, 42, 52, 54, 101–2, 270; of matter, 190; of parts, *see* differences; of relations, 27

equivalences: additive, 222, 234, 249; composition of, 219, 247, 253–4; co-ordination of, 203, 208, 217–19, 221; grouping of, 208–9, 224, 277; heterogeneous, 246; homogeneous, 246; intuitive, 220; operational, 247; perceptive, 166; qualitative, 183; simple, 220, 234, 238, 249, 255–6, 266; of volume, 232–68; of weight, 203–8, 210–16, 222–3, 227–8

experiments, *see* compositions; compression; conservation; density; displacement; equivalences; seriation

Gonseth, 228*n*

grouping: of classes, 170, 224; of equivalences, 208–9, 224, 277; of operations, 33, 61, 92, 103, 111, 115, 132, 267–73, 279; of relations, 33, 38, 42, 170, 201; of reversible compositions, 80

identification, 57–61

induction: experimental, 226–7, 231–3, 249–50, 267–8, 278; intuitive, 197

inequalities, qualitative, 183

Ingol-Favroz-Coune, Mme, 203*n*

intuition: and conflict with logic, 12, 212; and equivalences, 220; and evaluation of density, 143, 151; and evaluation of matter, 123; and evaluation of volume, 127; and evaluation of weight, 29, 158, 162–3, 166, 208; and induction, 197; and invariance, 60; and non-conservation, 270–1; and seriation, 197, 199; *see also* compositions; conservation

invariance, *see* conservation

invariants, 227–30

inversion, 157, 162, 176–7, 272

irreversibility: of actions, 12, 278; of compositions, 32; logical, 33, 90; of operations, 179; of perception, 271; of relations, 78–9; of transformations, 130; *see also* reversibility

Kant, 18, 269

Köhler, 269

Kowalewski, 270*n*

Lalande, A., 265*n*

length: seriation of, 188, 200, 220; and weight, 175–7

logic, *see* compositions; intuition; irreversibility; operations

mathematics, *see* conservation; operations; quantification matter: equalization of, 190; quantity of, 154, 159–70, 176–7; tangible, 279; *see also* substance
memory, tactilo-motor, 189
motivity, 279
movement, and weight, 33–5, 44

non-conservation, *see* intuition; substance; volume; weight

operations: compositions of, 89; co-ordination of, 17; direct, 271–2; equivalences of, 247; and experience, 232, 264–5; grouping of, 33, 61, 92, 96, 103, 111, 115, 132, 267–73, 279; identical, 272; intellectual, 200; inverse, 272; irreversibility of, 179; logical, 59, 77, 94–5, 155, 158–62, 172–3, 271–4; logico-arithmetical, 59, 102–4, 178–80, 199–203, 219, 225–31; mathematical, 59, 103, 114–15, 199–200, 273–4; physical, 59–61, 67, 87, 92, 95, 103, 110, 115, 131, 142, 158–62, 172–3, 178–80, 199–202, 217, 225–31, 274, 277; quantification of, 77, 177; of relations, 17–18, 187, 259–68; reversibility of, 9, 11–12, 17, 39–41, 44, 60, 77, 90, 110, 116, 130, 179, 225, 230–1, 259–68, 271–2, 278; and seriation, 184–5, 193–8, 277; of substance, 212–13; synchronisation of, 228–33; of weight, 212–13
Osman, Kiazim, 22*n*, 266

phenomenalism, *see* egocentrism
Piaget, J., 47*n*, 61*n*, 104*n*, 150*n*, 196*n*, 233*n*; and Szeminska, A., vii, 5*n*, 9*n*, 14*n*, 20*n*, 26*n*, 139*n*, 172*n*, 179*n*, 188*n*, 198, 209*n*, 213*n*, 271, 275*n*, 276*n*
Poincaré, H., 62
Proceedings of the Société de Physique et d'Histoire Naturelle de Genève, 20*n*, 162*n*, 255*n*, 264*n*, 276*n*

qualitative: equivalences, 183; inequalities, 183; reasoning, 85; relations, 44, 269
quantification: in compositions, 103; defined, 18; of density, 148; extensive, 18–19, 30, 38, 95, 163, 190, 204, 221, 270, 273–5, 278; intensive, 18–19, 29, 163, 221, 269–70, 274–5; mathematical, 18, 27, 204, 270, 274–7; metric, *see* mathematical; operational, 77, 177; of qualities, 16; of relations, 26, 38, 44, 269–78; of substance, 26, 45, 96, 113, 151, 154, 158–63, 170–2, 213–14, 275; of volume, 54, 96, 102, 113–14, 158, 161, 178; of weight, 29–30, 96, 102, 113–14, 152, 158, 161, 163, 170, 178, 210, 213–14, 217, 275; *see also* conservation

relations: asymmetrical, 183, 197–8, 208, 222, 225, 276–7; binary, 187; compositions of, 102; co-ordination of, 20, 41, 51–2, 55, 76, 114–15; equalization of, 27; grouping of, 33, 38, 42, 170, 201; inverse, 162; irreversibility of, 78–9; multiplication of, 155, 157; operational, 17, 18, 187, 259–68; perceptive, 17, 186–7, 199, 200; qualitative, 44, 269; quantitative, 26, 38, 269–78; seriation of, 225; spatio-temporal, 229
reversibility: of compositions, 39, 54, 80, 131–41, 220–1, 257, 260, 263; empirical, 134; operational, 9, 11–12, 17, 39–41, 44, 60, 77, 90, 110, 116, 130, 179, 225, 230–1, 259–68, 271–2, 278; *see also* conservation

seriation: compositions of, 205, 208, 213, 222, 225; co-ordination in, 192, 194; and egocentrism, 200–1; empirical, 190, 193, 197; intuitive, 197, 199; of length, 188, 200, 220; operational, 184–5, 193–8, 277;

pseudo- , 189–90; of relations, 225; simple, 188–91, 193; stages in 68, 105–8, 111, 186–98; of volume, 201; of weight, 183–202

Simon, T., *see* Binet, A. and Simon, T.

spatial relationships, 19

spatio-temporal: compositions, 94; relations, 229

Stage I in child development: in compositions, 185–90, 234–43, 261–7; in compression and density, 120–1, 130, 133, 137–40, 151, 156–9, 164–6, 175–6; in conservation, 5, 8–11, 17, 34, 45, 48, 60, 62; in conservation and atomism, 68–79; in seriation, 186–90.

Stage II in child development: in compositions, 165, 190–3, 234, 243–50, 261–2, 267; in compression and density, 120–4, 130–3, 137, 140–4, 151–2, 159–63, 167–77; in conservation and atomism, 61, 62, 68, 81–95, 101, 104; in seriation, 190–3

Stage III in child development: in compositions, 165, 193–8, 224, 250–7, 261, 267; in compression and density, 120, 124–7, 132, 134, 137, 144–8, 152–3, 173–7; in conservation, 5, 9–17, 22–35, 38, 45, 49; in conservation and atomism, 68, 98–9, 100–4; in seriation, 193–8

Stage IV in child development: in compositions, 235, 257–9, 261, 267; in compression and density, 120, 124, 127–37, 148–53; in conservation, 5, 22, 35–45, 53, 56, 61–3; in seriation, 68, 105–8, 111

Strauss, Trude, 154

substance: compositions of, 126, 146, 152, 177, 262–3; conservation of, 3–10, 12–27, 42–6, 52, 68–9, 81–96, 104, 120, 122–4, 133–4, 151; and density, 139, 142–3, 151–3; invariance of, 26, 45; non-conservation of, 69–79, 120–1, 133; operational, 212–13; permanence of,

39–40; quantification of, 26, 95–6, 151, 154, 158–72, 176–7, 213–14, 275; and volume and weight, 60–3, 74, 88–9, 111–14, 130–5, 140–68, 174, 177, 211, 214, 233–8

succession, law of, 189

synchronization, of operations, 228–33

time-lag, in child's development, 26, 55, 61, 178, 199–202, 227–8, 258, 261–2

transitions, 245

volume: compositions of, 129, 135, 171, 233, 239, 248, 260–8; conservation of, 3–5, 13–14, 47–63, 92, 97, 110–12, 127, 132, 177, 233, 275; and density, 137–43, 148–53, 155; displacement of, 5, 47–56, 232–68; equivalent, 232–68; increase in, 120, 124, 129, 131, 134–5; and intuition, 127; invariance of, 48, 94–6; non-conservation of, 51–5, 71–5, 94–7, 103–10, 132, 151, 170; quantification of, 95–6, 158, 161, 178; seriation of, 201; and substance and weight, 47, 60–3, 74, 88–9, 111–14, 130–5, 140–68, 174, 177, 211, 233–47, 265

von Weizsacker, 279*n*

Weierstrass, 270*n*

weight: compositions of, 126, 146, 152, 171, 203–31, 250, 261–5, 275; conservation of, 3–5, 12–14, 22–46, 68, 87–97, 102–3, 112, 126–7, 131–4, 152; and density, 137, 139, 142–3, 151–2, 170; equivalences of, 203–7, 210–16, 222–3, 227–8; and intuition, 29, 158, 162–3, 166, 208; and length, 175–7; and movement, 33–5, 44; non-conservation of, 7–8, 14, 26–40, 71–5, 94–6, 120–4, 127, 133–4, 142, 151, 170; operational, 212–13; quantification of, 26–33, 95–6, 158–63, 178,

213–14, 217, 275; and quantity of matter, 154, 159–76, *see also* substance; seriation of, 183–202; and substance, 60, 63, 74, 88, 111–12, 114, 144, 146, 151, 152, 211, 237–8; and volume, 147–68, 174, 177, 233–4, 237–8, 240–7, 265